FIREARMS

FIREARMS

THE LIFE STORY
OF A TECHNOLOGY

Roger Pauly

The Johns Hopkins University Press
Baltimore

First published in 2004 by Greenwood Press

Johns Hopkins Paperback edition, 2008
9 8 7 6 5 4 3 2 1

The Johns Hopkins University Press
2715 North Charles Street
Baltimore, Maryland 21218-4363
www.press.jhu.edu

Library of Congress Control Number: 2007940414

ISBN 13: 978-0-8018-8836-6
ISBN 10: 0-8018-8836-0

A catalog record for this book is available from the British Library.

Contents

Preface

Technological devices and systems are sometimes described as impersonal or inhuman. These are certainly apt phrases given the fact that it is often fixtures of cold metal components or collections of insensible wiring that one is forced to consider when studying such topics. Recognizing the soulless character of these utensils, it must still be admitted that some level of personification is often applied to them. Thus, a particular boat or a ship might be given a feminine moniker. If one takes an analogy like this a step further and applies it to a mechanical system as a whole, interesting and new perspectives emerge.

This book is one of the first in a series of Technographies produced by Greenwood Publishing Group. These works use the existing model of a biography and apply it to different technologies. While seeking to avoid overanimating mechanical devices, the series offers an innovative way to explore the holistic backgrounds of critical inventions. The subject of this particular experiment is the firearm, and the technographical approach delivers much ammunition in this particular case. The book shows how the development of the gun was a gradual and uneven process much like the growth of an individual. Common imperatives drove a sporadic evolution as this weapon passed through many hands and through many years.

Thanks goes out in particular to Kevin Downing of Greenwood Publishing, the editor of this series and the person who first developed the concept of the technography. Gratitude must also be given to my colleagues at the University of Central Arkansas, foremostly Ron Fritze who has acted as an invaluable mentor on this project and in many other ways. Lastly, some recognition should be given to my family, who have had to put up with the demands of the writing schedule along with the usual ranting and ravings.

Introduction: Pandora's Barrel

In the mid-nineteenth century, Captain Augustus Henry Lane-Fox was part of a committee charged with finding a replacement for the venerable Brown Bess musket that had served the British Army for well over 100 years. As part of his work Lane-Fox had to look at a wide variety of contemporary and older firearms, and he was struck by similar patterns in design. A decade or so later, he applied Charles Darwin's new model of natural selection to weapons and began to argue that technologies can evolve in a manner similar to biological species. Eventually Lane-Fox included a display at an anthropological museum he founded in Oxford that demonstrated how early members of the firearm family had mutated into higher, more efficient forms.

While Lane-Fox's analysis of technological development had merit and was shared by others in the emerging field of anthropology, it by no means exhausted the potential models available through which one can analyze a technology. Rather than considering guns in the manner of biological individuals, one might rather approach the topic as if it were a singular being. In other words, the millions of individual weapons consisting of thousands of models and designs might well be considered representatives of some archetypical entity known simply as "the firearm." Without excessive personification, this work will seek to employ such methodology in order to offer a short, manageable review of the history of this important technology.

When looking at the development of firearms, it is possible to make some analogies to a biography of an individual human being. Thus, early human missile weapons like spears, slings, and bows represent the distant ancestors of the gun. The crossbow could probably be identified perhaps as a more immediate parent, with its telltale lock mechanism and its stock-like tiller. Another parent might well be a medieval Chinese flamethrower known as a "firelance" that used gunpowder as an incendiary. The earliest hand-cannons of Asia and Europe illustrate the gun's childhood, while the diverse weaponry of the sixteenth and seventeenth centuries is perhaps reflective of the rapid developments of adolescence. In the nineteenth century new industrial techniques allowed the firearm to surge into a stage of adulthood as metallic cartridges, revolvers, breechloaders, and repeating rifles appeared. On the eve of the twentieth century, the recognition that gas and recoil from a shot could be used to drive a gun's reloading mechanism resulted in the rise of both semi- and fully automatic weapons. These last categories dominate firearm designs today. While admitting that lighter and more precise weapons continue to emerge, the major operating systems have not changed drastically in recent decades, and thus one might conclude that the firearm had reached a period of maturity if not old age.

Like a human, this growth was uneven and erratic. As outlined above it was not marked by steady progress, but rather through periods of intense development and changes followed by episodes of slowing or stagnation. Often future gun designs were anticipated well in advance of the necessary technological maturation, much like a child dreams about a future career well before she is capable of piloting a jet fighter. This led to some cases in which an individual like Samuel Colt has erroneously been credited as an "inventor" of a gun design when a title like "operational developer" might be more appropriate. In a few cases, an Alexander John Forsyth or a John Moses Browning actually created something new. For the most part, however, gun designers often just improved earlier systems and made them more functional.

It is also important to keep in mind that the firearm was the creation of human beings and thus cannot be divorced from society. In most cases, it was the deadly serious business of warfare that drove the firearm into more lethal forms. Even by the Middle Ages, far more people around the globe obtained their food through agriculture, not by means of hunting. The latter was primarily done for sport, and as such it was not nearly as much of a priority for weapons development as was combat. In the latter realm, very often the firearm was changed by demands coming from the battlefield. The reverse case also happened, and certain guns came to drastically change the battlefield itself.

While the whole life story of the firearm might appear erratic and unpredictable, some consistent logic lay behind most of the process. No matter what type of missile weapon one is considering, it has to address certain demands or meet certain needs. First of all, it must be able to hit its target. An exceptionally powerful gun will generally have no effect on an enemy if its bullet does not actually hit him or her. Toward this end, small arms manufacturers have consistently concerned themselves with promoting a firearm's accuracy. Linked to this to some extent is the importance of range—in other words, the potential to hit objects at relatively far distances. On the other hand, it does no good to consistently hit an enemy or intended meal with a projectile that does not actually inflict any damage upon them. Therefore, the power and strength with which a projectile crashes into its target are important too. A third ingredient to these two factors is rate of fire. Poor accuracy or weak power can be offset to some extent with a weapon that discharges many shots in a short period of time. Being fired at by a single powerful, but highly inaccurate, smoothbore musket just once may not prove lethal. The situation becomes drastically more serious if an individual is shot a number of times by several dozen muskets. Thus the common imperatives of accuracy, power, and rate of fire guided the development of the firearm at each stage of its life cycle.

It probably should also be noted that unlike many technologies, the firearm has played a somewhat villainous role in history. Although some inventors tried to claim otherwise, gun innovations were made largely in order to kill people more efficiently and effectively. One can emphasize the noble ends or purposes that have forced good people to take up guns in defense of their persons or liberties, but this does not mitigate the nefarious intent for which these weapons were created. The same types of firearms that have been used to protect "democracy" or "freedom" have also been used to commit absolutely horrific crimes. On the other hand, one also has to admit that there is something of a love affair between humans and these destructive devices. The same celebrity who might publicly denounce guns and youth violence will apparently think nothing about making a movie that glorifies them to a young audience. There is an intangible effect in the report, flash, recoil, and obvious power of a discharging firearm that attracts people of many backgrounds and temperaments. For lack of a more eloquent phrase, it can simply be noted that guns can sometimes seem fun.

It is difficult to balance the dubious function of the firearm with the human fascination over it. The paradox is reminiscent of the ancient Greek myth involving Pandora's box. In this tale the first woman ever created was forced to balance a warning from the gods against her own curiosity. Although she was cautioned to not open a mysterious container, she eventually unfastened its lid to see what was inside. The gods had hidden all the woes of the world in the box, and Pandora unwittingly released them. Once freed, they could not be put back in and humanity was forced to suffer with the consequences. Likewise, on each occasion that an innovation or improvement in the firearm was made and became commonly known, the metaphorical bullet could not be put back into the gun's barrel. The technology spread, was copied, and was adapted to new situations. In this way, for better or for worse, the firearm has become a key human implement. Like the broader story of humanity itself, the biography of this weapon has been marked by fortune against failure, innovation following imitation, and success mixed with stagnation.

Timeline

Prehistory	Common missile weapons such as spears, javelins, atlatls, slings, and bows in use.
200 BCE	Relatively sophisticated crossbows in use in China.
Late first millennium	Early flammable powders experimented with in China.
Around 950 CE	Possible artistic depiction of an early firelance.
1040 CE	Oldest printed recipe for gunpowder in China.
1128	Buddhist wall carving possibly depicts an early gun.
1280	Islamic work detailing seventy different gunpowder formulas printed.
1288	What may be the world's oldest surviving gun dropped on a battlefield in Manchuria.
1326	First printed references to cannons in Europe.
1332	World's oldest surviving definitively dated gun cast in China.
Around 1350	Possible date of the casting of Europe's oldest surviving gun, later found near Loshult in Sweden.

1390s–early 1400s	Several surviving specimens of early "handcannons" are built, including one with four barrels.
1400	Increasing use of "handcannons" in Europe.
1410–1440	Matchcord and serpentine ignition systems developed.
1471	Twenty percent of Burgundian army's missile-firing troops use firearms.
1490	Venice begins replacing all crossbowmen with handgunners.
Around 1480–1500	First use of rifling in gun barrels.
Early 1500s	Ottoman Turks wage successful campaigns against Mamluk and Persian neighbors by using firearms.
Around 1490–1519	Leonardo da Vinci sketches image of a wheel lock.
1515	First written reference to a wheel lock.
1522	Spanish troops equipped with matchlock arquebuses inflict defeat on Swiss pikemen at Bicocca.
1524	Famous knight, Pierre du Terrail, chevalier de Bayard, shot dead by common musketeer.
1526–1527	Mughal ruler Babur wins key victories at Panipat and Kanua in India using guns.
1537	Early use of arquebus-armed dismounted cavalry as mobile infantry.
1540s	A matchlock with three revolving barrels is built in Venice.
1549	Japanese leader Odo Nobunaga begins equipping his troops with domestic copies of Portuguese arquebuses.
Around 1550	Development of snaplocks in Europe.
1560s	A German wheel lock pistol with three locks designed to fire three separate shots from the same barrel is built.
1575	Odo Nobunaga wins decisive battle of Nagashino using firearms.
1597	Hans Stopler of Nuremberg builds an arquebus revolver.
1599	Maurice of Nassau standardizes the caliber sizes of his troops' guns.
Around 1600	Some Danish royal guard are issued rifled weapons.

1601	Most European infantrymen now carry firearms.
1610–1620	First "true" flintlocks developed in France.
Around 1650	Paper cartridges, developed decades earlier, start to become widely popular.
1689	The major powers of Europe begin switching to flintlock muskets as standard weapons for their infantry.
1718	James Puckle patents a revolving gun that anticipates elements of later manually operated machine guns.
1721	Isaac de le Chaumette patents a breech-loading rifle.
1730s–1750s	Hunting birds in flight becomes popular due to the development of lightweight shotguns.
1740s	Benjamin Robins determines the physical principles behind the accuracy of rifled firearms.
Around 1750	German immigrants in English North American colonies begin modifying short Jaeger firearms into longer Pennsylvania rifles.
1776	Patrick Ferguson demonstrates his breech-loading rifle based on Chaumette's early design to George III.
1790–1810	Pennsylvania rifles gradually evolve into even longer Kentucky rifles.
1799	Englishman Edward Howard unsuccessfully attempts to use mercury fulminate as a gunpowder.
1800	Ezekial Baker develops a short Jaeger-like rifle for the British army.
1803–1812	Russian army issues some 20,000 "Tula" rifles to its troops.
1807	Scottish clergyman Alexander Forsyth patents the first percussion lock after successfully employing mercury fulminate as a primer.
1811	American John Hancock Hall patents a breech-loading rifle and later perfects the manufacture of standardized parts.
1812	Swiss Samuel Johannes Pauly demonstrates a hinge-action firearm he developed that uses metallic self-primed cartridges.
1818	American Elisha Collier patents a flintlock revolver pistol in Britain.

1820s	British Captain John Norton develops an experimental expanding rifle bullet.
1836	British gunsmith William Greener presents a round two-piece expanding rifle bullet.
1836	American Samuel Colt patents his percussion-lock revolver pistol.
1837	Russian poet Alexander Pushkin dies after being shot in a pistol duel.
1838	Flawed Brunswick rifle using "belted" ammunition introduced in Britain.
1838	Nicholas von Dreyse invents bolt-action "needle gun" in Prussia.
1848	American Walter Hunt patents his "rocketball" ammunition.
1849	French captain Claude-Etienne Minie presents a cylindroconoidal-shaped expanding rifle bullet that quickly becomes popular in many countries.
1850s	Robert Adams and Frederick Beaumont develop the double-action revolver in Britain.
1853	The British Enfield rifled musket is developed.
1853	American Christian Sharps perfects his breech-loading percussion carbine.
1855	The U.S. Army adopts a Springfield rifled musket using a taped primer ignition system developed earlier by Edward Mayner.
1857	Some Indian "Sepoy" troops mutiny against their British officers in part over their new Enfield rifles and accompanying Minie-designed ammunition.
1860	American Benjamin Tyler Henry (working for Oliver Winchester) and Christopher Spencer independently develop similar lever-action repeating rifles.
1861	The United States abandons the Maynard primer in a new version of the Springfield rifled musket.
1861	American Wilson Agar develops a single-barrelled, manually operated, "coffee mill" machine gun.

1862	Dr. Richard Gatling patents a more successful multibar-relled manually operated machine gun.
mid-1860s	Horace Smith and Daniel Wesson develop hinge-action "breech-loading" revolvers.
1866	Winchester develops the first of several improved versions of its original lever-action Henry rifle.
1867	The Jacob Snider breechloader conversion is adopted for the British Enfield while Russia selects a similar conversion for its Tula rifled muskets developed by Czech inventor, Sylvester Krnka.
1868	The United States begins developing new "trapdoor" breechloaders based upon conversions of older Springfield rifled-muskets.
1870	During the opening of the Franco-Prussian War, the French army fails to effectively deploy their manually op-erated "Mitrailleuse" machine guns developed by Josef Montigny.
1871	British Army adopts the breechloading Martini-Henry rifle, which uses a "Bottleneck" cartridge.
1873	The Colt "Peacemaker" revolver is introduced.
1874	France converts an earlier bolt-action breechloader devel-oped by Antoine Alphonse Chassepot to use metallic car-tridges.
1877	Turkish forces equipped with repeating rifles inflict severe losses on Russian troops at the siege of Plevna.
1884	French scientist Paul Vielle develops Powder B, a "smoke-less" propellant.
1885	Hiram Maxim presents his recoil-driven, water-cooled au-tomatic machine gun to the public in London.
1886	The French Army adopts the Lebel rifle, a bolt-action weapon utilizing Vielle's inventions.
1886	Austrian nobleman Ferdinand Ritter von Mannlicher devel-ops a bolt-action rifle using a fast "packet" or "clip" reload-ing system.
1888	Britain develops the Lee-Metford bolt-action rifle, which uses a detachable box magazine.

1889	Peter Paul Mauser develops a "charger" or "strip-clip" system, which is similar to that used in the Mannlicher but becomes much more popular.
1892	Austrian Joseph Laumann develops the "Schonberger," a semiautomatic pistol.
1893	Hugo Borchardt of Germany develops a semiautomatic pistol with a detachable box magazine hidden in the handle that would inspire the famous "Luger" and other similar handguns.
1895	Colt produces a gas-operated, air-cooled automatic machine gun invented by American John Moses Browning.
1896	Danish army captain W. O. Madsen develops a similar lightweight air-cooled, recoil-operated machine gun.
1900	Browning begins to design a popular series of semiautomatic pistols to be produced by Fabrique Nationale of Belgium.
1913	American Isaac Lewis develops a highly portable, air-cooled, gas-operated light machine gun.
1916	Inspired by a converted fully automatic Luger pistol, German Hugo Schmeisser develops the Bergman Machine Pistol, an exceptionally lightweight automatic weapon using pistol ammunition.
1917	Browning develops his "Browning Automatic Rifle," or BAR, a hybrid weapon capable of selective semiautomatic or fully automatic fire.
1919	Retired U.S. Army general John Thompson coins the phrase "submachine gun" to describe a weapon that he invented that was similar to the Bergman.
1936	The U.S. Army adopts the semiautomatic M-1 rifle developed by Canadian-born John Garand.
1943	A German team develops a light-caliber rifle capable of selective fire that will become known as the "Sturmgewehr," the world's first assault rifle.
1947	Soviet designer Mikhail Timofeevich Kalashnikov improves on the Sturmgewehr and creates the most popular firearm of all time, the AK47.
1963	The United States begins purchasing large numbers of Colt's light-caliber M-16 assault rifle.

1985   Britain adopts the troublesome "Bullpup"-style SA 80 as-
                        sault rifle.

2000   United States grants award to Alliant Techsystems to
                        develop the XM29 "Objective Individual Combat
                        Weapon," which is intended to become a combination
                        assault rifle/air-bursting grenade launcher.

1

Ancestors: The Ancient Origins of Firearms Technology

◆

PRIMORDIAL MISSILE WEAPONS

The noted historian Alfred Crosby recently observed that throwing things is an indelible element of human existence. In addition to walking upright and manipulating fire, he argued that the ability to effect change at a distance is one of the fundamental characteristics that separates Homo sapiens from other members of the natural order (Crosby 2002, 3–4). At one time, scholars made the broad argument that the very use of any tools whatsoever was a key distinction. The more recent recognition that otters, chimpanzees, and some orangutans employ rocks and sticks to obtain food has clouded this division with our furry utensil-wielding colleagues.

In any event, early humans more than compensated for their limited biological weaponry by engaging the services of implements. Equipped with basically flat teeth and a pathetic version of claws, humanity grasped for whatever inanimate objects could fulfill similar functions. At the most basic level, a stick or a rock could be employed to deliver a harder blow than a bare fist. The process by which this technique may have emerged was famously depicted in Stanley Kubrick's classic science-fiction film *2001: A Space Odyssey.* In this work, an early hominid learns to kill his enemies with a bone. The significance of such a moment in human history is portrayed as critical and apparently leads directly to space exploration. While this technological evolution takes place in the course of a scene change during the movie, one needs to take a more leisurely pace when studying the history of a technological system.

A number of more specialized tools had to be developed before the emergence of the firearm could occur. The gun had a long and distinguished family tree with many important ancestors. Not all of these show a direct connection with firearms, but the imperatives behind their creation remained constant. The idea was to deliver as much damage as possible while hopefully avoiding retaliation. There were two elements in this endeavor. Primarily, there was an effort behind many developments in missile technology to increase the force and the distance at which blows could be delivered. In addition, the actual number of shots that could be launched in a given time period was important too. If all of these factors could be increased, there would likewise be a corresponding rise in the potential level of damage inflicted upon either foe or food. Toward this end, accuracy was naturally an important factor as well.

Perhaps the best way to deal with an enemy armed with a bone, club, axe, or sword would be to prevent him or her from getting close enough to strike a blow in the first place. This is particularly useful if your opponent is of greater physical strength and would be likely to overwhelm you in hand-to-hand or melee combat. The first "missile weapons" were likely to have been hand-thrown rocks. In fact humans were probably chucking stones at hostile animals and each other long before anyone tried to tie one to a stick and make a club. As ancient as this method is, it remains effective. Riot police around the world today are routinely wounded by such simple methods. The thrown rock also has at least one clear advantage over other simple weapons systems: rate of fire. A single person with a good supply of stones can unload several dozen shots in a minute. This can be an exhausting effort, however, and sooner or later the speed of each stone will taper off. Furthermore, as many of us right-field veterans know, not everybody has "an arm" in the first place. Happily for such unfortunates, the chance to disrupt attacks from either fang-bearing critters or club-wielding bruisers could be magnified by simple technological implements.

Some of the developments in this regard evolved around the effort to extend arm reach and thus also centripetal force. The simple sling is one example of this. It is a clever device that can easily be manufactured from leather or cloth. The sling consists of a small receptacle or pocket with two cords tied to opposite ends. A projectile such as an ordinary rock, carved stone ball, or lead bullet is placed in the receptacle while one cord is fastened to the wrist or lower arm. The other cord is held in the hand. The loaded sling is then whipped at the target and, if done properly, the unfastened cord is released at a precise moment. While different throwing techniques or styles might be employed, the device effectually increased the thrower's arm length and the velocity of the flung object. The sling's extended cords added speed to a throw. Slingers were employed effectively throughout the ancient world and were still being used in European warfare at least as late as 1572 (Reid 1984, 17).

In addition to rocks, pointed sticks certainly represented another primordial missile weapon. To add forward weight to a wooden rod and improve its aerodynamics, it was usually fitted with a pointed spearhead. These were initially made of stone, but copper, bronze, iron, and steel would eventually be employed too. Beyond the aerodynamic value of a weighted front, the spearhead had the additional responsibility of cutting through hide, skin, and armor. They would be filed and sharpened in order to achieve this effect. The spear also offered the side benefit of being effective in combat even when *not* thrown. Eventually very heavy thrusting spears and lances were developed for this purpose alone. In general, however, the original idea behind the spear was to inflict long-range damage, and the atlatl was an aid toward that end.

The atlatl manipulated the same physical principles behind the sling. It is another simple device consisting of a 1–2-foot-long stick designed to hold a spear or long dart. The atlatl typically had a carved groove in which the shaft of the missile rested, while its blunt end nestled in a hook or a pocket. Instead of grasping the spear directly, the thrower held the "loaded" atlatl and swung it in the direction of the target. Like the sling-thrown rock, the missile was effectually launched as if thrown by someone with a phenomenally long arm. A noted Spanish conquistador, Bernal Diaz, recounted the fast rate of fire and lethal effects of these weapons. In his description of one battle in Mexico, he described how "fire hardened darts fell like corn on the threshing floor, each one capable of piercing any armour" (Diaz 1963, 149). The very term "atlatl" is in fact Aztec in origin, although other peoples have used the same basic devices. There were a number of variations on both the sling and the atlatl, and many of these have continued to be employed here and there up through modern times. These missile weapons are a bit awkward to use, however, and require a lot of practice in order to achieve much accuracy. Other ancestors in the firearm's family tree managed to better address these limitations.

One of the more remarkable early projectile weapons did not rely upon arm power at all but rather upon the force of one's lungs. Those who have ever shot a spitball out of a straw have some familiarity with the principles behind this device. The first blowguns were probably made of hollowed-out pieces of bamboo or reed. A small dart could be propelled by lungpower through the tube and out at an enemy. This weapon remained popular with certain nonindustrialized groups well into the twentieth century, but it did have a very limited range and little stopping power. Blowgun darts often had to be treated with poisons to make them effective. The weapon is also of interest to our story, however, because of its physical similarity to the firearm. The idea of forcing a projectile down a tube clearly lies behind both weapons. Air guns certainly take advantage of this principle, but it is less clear, however, if the firearm really owes its direct lineage to the blowgun. The device does not seem to have enjoyed particularly widespread usage in comparison to other missile weapons.

UNIVERSAL PROTO-FIREARMS: BOW WEAPONS

The same certainly cannot be said about the arrow-firing bow, an extraordinarily successful and long-lived technology. It found employment with a vast and diverse clientele stretching across the continents and the centuries. Warriors of ancient Mesopotamia were using essentially the same device that many peoples are still employing recreationally and otherwise today. Benjamin Franklin, one of the most noted scientific figures of his time, suggested that the American forces consider abandoning their muskets in favor of the fast-firing, easily produced bow. Indeed, the bow and arrow combination was so successful that it is not entirely clear why early firearms were ever developed in the first place. One series of experiments conducted recently suggests that a simple bow was more accurate and had as much as twelve times the rate of fire of an eighteenth-century musket (Given 1994, 100–110).

The ever-present stick was likewise used in the construction of this device. In its simplest manifestation, the bow consisted of a 3–6-foot-long straight staff, preferably made of an exceptionally tough wood such as yew. A somewhat shorter cord was fastened to both ends, pulling the entire device into the shape of a letter "D." The arrows that were launched from the bow were likewise made of straight sticks, although these were generally shorter and narrower in diameter. They also typically carried small arrowheads for the same flying-and-cutting purposes behind the spearhead. As an additional measure to improve accuracy, the rear end of the arrow was often fitted with feathers that stabilized flight by creating rotation. A notch was usually carved into the wood at the very base of the arrow in order to help it "grasp" the bowstring.

The forces being manipulated in the bow differed from those employed in the sling or atlatl. The power behind the arrow is the result of a sudden release of energy created by the bow trying to resume its original shape. This tension is first produced when the archer strung the normally straight bow, as noted above. The overall pressure on the shaft was considerably increased through the action of fitting an arrow to the string and pulling both back. At the moment the archer chose to release her grip on the string and arrow, the shaft of the bow would suddenly snap as far back to its natural shape as the string would allow. Assuming the bowman did not get his other hand, the shaft, his hair, or any of his clothing in the way, the arrow would be forcefully flung forward through the air toward his target.

The advantages of this system over the sling or the atlatl are fairly clear. Looking again at the universal imperatives of force, range, and effective number of shots, the bow seems to be the superior device. A sling or a javelin can certainly hit with a high velocity, but bow-fired arrows are not lacking anything in this regard. As far as distance goes, in the same person's hands, the bow and arrow will usually outrange the slingshot rock or atlatl-hurled javelin. A primitive atlatl might hurtle a javelin 100 yards, while a sling could possibly hit an enemy at twice that range. A basic wooden bow could equal the latter shot, while more specially designed ones could greatly exceed it. English medieval archers equipped with longbows could routinely fire at ranges of 300 yards (Hardy 1992, 54). In the late eighteenth century, Turkish archers were known to have launched specially designed arrows over a half of a mile! In practical terms the bow also offered a high rate of fire. A good archer could fire ten shots in a minute. While a shot every six seconds is probably beyond the ability of the average person, it can be admitted that the bow and arrow combination had a higher overall rate of fire than the atlatl. Arrows are generally a bit smaller and lighter than javelins, and an archer can thus carry more projectiles. On the other hand, one can likewise carry many more small stones or bullets than arrows. Nevertheless, the number of *effective* shots the average archer made probably remained higher than that of the average slinger. The manner in which the arrow was drawn created a clear line from the archer's face and eyes toward his other arm and the point of the arrowhead. Targets could be sighted along this line with a fair degree of ease. It took quite a bit more practice to achieve a comparable level of accuracy with a sling. Whipping a sling or an atlatl also required an exhaustive, full-body exertion, similar to what a major league pitcher does during a windup. Although the pitcher is well compensated, this can be grueling work, as the existence of a bullpen will indicate. Drawing back a heavy bow was by no means easy, but it did use a comparatively fewer number of muscles. It is also somewhat easier to fire a bow from behind a tree or the corner of a building.

The basic design of the bow was often tinkered with and improved in order to make the device more powerful. The shaft might be made of a composite of materials including wood, bone, and/or horn, which were fitted together to make a stronger and more durable device. Rather than starting out as a straight staff, some bows were designed in a reverse "C" with the ends bending far forward when unstrung. This "reflex" bow could thus pack a lot of punch in a more compact package. Because the reflex bow was made shorter from end to end, it could be used effectively on horseback. It is no surprise therefore that this fancy-looking but deadly device was employed by a number of cultures in Central Asia that favored horseback riding, such as the Mongols. With the help of this weapon, the Mongols carved out an empire stretching from Korea through Baghdad to the borders of Hungary. The footslogging English bowmen of the Hundred Years' War had no apparent need for such advanced designs, and continued to employ an older, straight-shaft version known as the long bow. A century after the conquests of the Mongols, these archers managed to mow down expensively armored French knights with this incredibly simple device, little altered from Neolithic times.

Given this success, Franklin's commentary seems highly insightful. An old saying comes to mind: "If it ain't broke, don't fix it." The fact that someone had to invent this saying at all suggests something about human nature. People *do* try to fix things that work perfectly well, just to explore the possibilities. Those of us who routinely have to suffer through the process of getting computers "upgraded" every other year can attest to the persistence of this annoying tendency amongst our fellow humans. In the case of the bow, an alternative appeared in Asia that, at first glance, does not seem to have conveyed an immediate advantage.

As the second syllable in its name suggests, the crossbow is just another variation of the bow-and-arrow system. It utilized the same basic principle to forcefully propel a projectile, only in this version the bow was fixed perpendicularly astride a 2–3-foot-long wooden stock known as a "tiller." The bow component lay across one end of the tiller, hence the term *cross*bow. From above, the device actually looks less like a traditional cross and more like a letter "T." The exact date of its invention is lost to time, but fairly refined models were already being used by the end of the third century BCE in China.

Unlike the missile weapons reviewed above, the crossbow was not an intrinsically simple device, and it demanded a certain level of craftsmanship to produce. In the basic crossbow, the trickiest component was the lock. This bronze, iron, or steel mechanism was fitted into the tiller, on the end opposite from the bow. To fire the weapon, the string (really more of a cord in this case) was pulled back and held in a fixed position by the lock. A short version of the arrow, known as a bolt, was then placed in a groove carved along the top of the tiller. Just as the grooved atlatl held the dart, so did the grooved tiller secure the bolt, the back of which butted up to the drawn-and-locked cord. In later variants, the bolt was held more firmly in place by a springlike piece of horn or steel. In addition to its diminutive length, the typical bolt was also a bit thicker than the standard arrow and substituted wooden fins for feathers. The lock mechanism could be released suddenly by means of a trigger that extended out from the bottom of the tiller. When this happened, the cord was snapped forward, driving the bolt along its groove and sending it flying off through space.

Over time, the basic crossbow underwent a number of improvements. Originally the cord was pulled back (or "spanned") by hand. As bows were made stronger, increasingly complicated means had to be devised to enable them to be loaded. The crossbow was fitted with a stirruplike device at its firing end, while the archer wore a specially designed spanning belt equipped with a hook. The unloaded bow was set on the ground, firing end forward, with the soldier's foot planted firmly in the stirrup. He then bent over the bow and attached his hook to the cord. By pulling himself up straight, the cord was drawn back into the lock. Toward the later part of the Middle Ages in Europe, even more powerful variants with steel bows began to be constructed. This added considerably to the velocity of the bolts it fired, but it also added considerably to the strength required to span it. To accomplish this, a number of specially designed cranks had to be devised that could be fitted to the tiller. There were variations, of course, but the most common types were the two-handed "windlass" and the one-handed "cranequin." The former of these fit on the rear of the tiller, while the latter hooked to the side. These innovations certainly improved the force and impact of the average crossbow's dart over that of the average bow-fired arrow.

On the other hand, they clearly could *not* have improved the weapon's rate of fire. In the Hundred Years' War, English longbowmen generally bested the crossbowmen used by the French. While there may have been a number of reasons for this, the longbow's much greater rate of fire was very likely a factor. Even at *five* shots a minute, the longbow would certainly send far more projectiles at an enemy than a cranequin-assisted crossbow. Remember that the basic demands behind missile weapon technology have remained relatively constant throughout history. Humans were invariably trying to increase the force, the range, and the number of missiles hurled at a target. While the crossbow offered an improvement in power, it was clearly inferior in the quantity of projectiles it could deliver.

A medieval crossbow being spanned: the parent of the firearm? © Hulton Archive/Getty Images.

The crossbow's low rate of fire helps explain the continued widespread persistence of the bow and arrow, even in relatively technically adept societies like late medieval England. Still, quick reloading was not paramount in every situation. Hunting remained an important use of projectile technology well past the period of the Middle Ages. Even a 6-second time period is too long to reload when hunting a swift deer or an aggressive boar. The crossbow's harder hitting bolt conveyed an advantage in this realm, and sporting versions of the weapon were used right up to the early nineteenth century. In fact, for that matter, specially designed crossbows made with modern materials are still used for this purpose today. The idea behind modern archery hunting may be to get back to basics, but somehow compound bows with pulleys and crossbows equipped with telescopic sights don't seem quite appropriate in this regard.

In a roundabout way, the crossbow actually did offer a better effective rate of fire for some people. An English longbow might have a "pull" of 100 pounds or more. In other words, pulling back the bow would be the equivalence of lifting a similar amount of weight. Although few men might *admit* this is difficult, it is. In fact, a lot of people would have been physically incapable of firing such a weapon. In their hands, the longbow had an extremely low rate of fire: none a minute. A cranequin-equipped crossbow, however, allowed weaklings to shoot a projectile with a power similar to that of a muscular bowman. The number of *effective* shots again comes into issue here. It is fatiguing business to hold a taut bowstring while aiming. Good archers learned to pull, aim, and fire quickly. The crossbow's lock, however, allowed one to take all day carefully lining up a shot if so desired. While the survival of the basic bow and arrow is understandable, it is likewise clear why the crossbow developed the popularity that it did.

A particularly notable feature of the crossbow is its physical similarity with later firearms. Even the earliest medieval crossbow shares some obvious characteristics with the later gun, particularly in the stock and lock. Remove the bow and you have two of the three major components of later muskets and rifles. The only thing missing would be the tubelike barrel. In a paternity lawsuit based on appearance alone, the crossbow could make a strong claim as a parent of the firearm.

GUNPOWDER AND THE FIRELANCE

At first glance, the advantages offered by early firearms themselves seem even less obvious than those conferred by the crossbow. It is not at all certain that early handguns provided a clear improvement in terms of velocity, range, rate of fire, or accuracy. In fact the birth of the handgun may have been more directly related to a very different weapon entirely. Ultimately the firearm was a by-product of a chemical invention. This mixture is perhaps best called gunpowder in reference to its most famous function. The term "blackpowder," a commentary on its appearance, is believed to be a later appellation.

While gunpowder was primarily a Chinese innovation, it may have received some Indian inspiration. Just as China embraced Indian Buddhism, the subcontinent's fascination with fire may have likewise crossed the Himalayas. In 664 an Indian visitor to China reportedly demonstrated the peculiar flammability of saltpeter and provided instructions on how to locate it (Pacey 1990, 16). It is at least equally possible, however, that Taoist alchemists in China, long interested in the natural order of the world, had already come across it. The term "saltpeter," incidentally, is a bit misleading. Although it is commonly associated today with potassium nitrate, terms like "saltpeter" have been used to designate a variety of nitrate salts. Different types of nitrates have been used in the production of gunpowder, with the result that performance could vary quite a bit.

Regardless of whether the saltpeter technology was transferred from India or not, the further development of gunpowder was clearly a Chinese accomplishment. It is less clear, however, how intentional this invention was. It may be that the alchemists systematically experimented with saltpeter and with more common flammable substances like sulphur and charcoal. Early written evidence from the mid-ninth century suggests, however, that some of these discoveries were accidental. Taoist texts specifically warn their readers *against* making mixtures containing such volatile elements due to past mishaps (Needham 1986, 111–113).

In spite of such warnings (or perhaps *because* of them), individuals continued to experiment and had probably figured out the basic formula of saltpeter, sulphur, and charcoal by about 950. The latter two ingredients were not hard to come by. Sulphur is a common mineral that is much more readily available than saltpeter. The Taoists were able to find it with relative ease due to its characteristic yellow color. Carbon was obtained from charcoal, a simple by-product of partially burned wood. The Chinese alchemists had discovered that these three flammable substances became even more combustible when mixed together. They burned at an extraordinarily rapid rate in a sudden "poof!" complete with a lot of smoke, a strong smell, and a bright flash.

The earliest actual printed recipe for this "fire chemical" appeared in China about a century later, around 1040. Nevertheless, several centuries of tinkering remained in order to perfect the proper percentage of each ingredient in the mixture. The earliest versions of gunpowder were relatively weak in power and uneven in temperament. Gradually the proportion of nitrate in the formulas rose, along with their relative strength. The most effective recipes for gunpowder eventually suggested a ratio of about 75 percent nitrate and roughly equal remaining measures of charcoal and sulphur in order to produce the most powerful reaction (Partington 1999, 324–328). It was also believed that certain mixtures were better for some tasks than others, depending upon whether one wished to launch a rocket or fire a cannon. An Islamic work dating to 1280, a mere century or so after the invention is presumed to have spread beyond China's borders, offers over seventy different recipes (Diamond 1997, 247).

Gunpowder was certainly not the only spectacular combustible concoction in the world at this time. Flammable oils, resins, and petroleum products were also being experimented with in a number of countries. A nasty, napalmlike Byzantine mixture known as "Greek fire" was perhaps the most flamboyant example of this. Nevertheless the black, dry, combustible compound that the Chinese created was the key chemical invention behind firearms for several reasons.

Fire is a chemical reaction that is dependent upon the presence of three things: oxygen, fuel, and heat. In this process, vaporized fuel reacts with oxygen and gives off light and even more heat. Charcoal and sulphur thus need oxygen in order to burn. Nitrates, on the other hand, contain oxygen molecules bound within them that can provide for the fire's needs. This means that a substance like potassium nitrate will burn within a confined area, without access to outside air. Fuel and oxygen, two of the factors needed for fire, exist within the gunpowder itself. Gunpowder could thus be packed into confined spaces and, if provided with sufficient heat, still ignite. This is one critical attribute of gunpowder.

Fire is often viewed as a force of destruction, and this is rightly so, given the observable end effects of weapons that employ it. From a chemical standpoint, however, fire is more of an agent of transformation. Matter can be transformed, but in an absolute sense, it cannot be destroyed. In a conflagration, fuel is changed into vapor, heat, and light. The basic number of atoms existing before and after the reaction remains the same.

One of the more remarkable features of burning gunpowder is the manner in which it transforms into vapor. When ignited, this peculiar mixture creates an enormous expansion of gases, on the order of 6–7 gallons of vapor for each 30 grams or so of solid powder burned. The rapid expansion of those gases represents an enormous potential force that could be directed toward a particular task. The effort to better drive this power

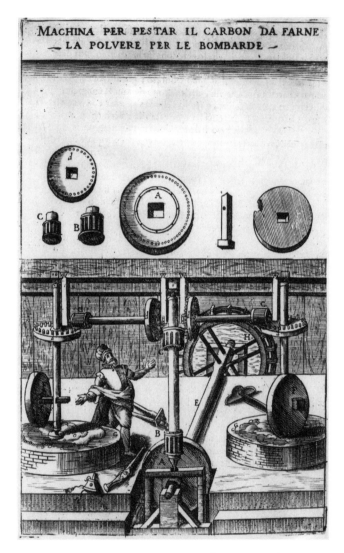

An early modern European gunpowder factory. © Hulton Archive/Getty Images.

against an opponent or potential meal represents much of the history of the firearm. This expansive force is a second key attribute of gunpowder.

Technically speaking, the mixture creates an extremely fast burn known as a "deflagration." The shock wave from a deflagration explosion travels slower than the speed of sound. This is to be distinguished from a "detonation," which has a supersonic shock wave. The invention of detonation-producing explosives would not happen until centuries later.

Such subtle differences meant little to those experimenting with early gunpowders. For all practical purposes, gunpowder did in fact explode. The stuff was fascinating, even if its shock wave only traveled at less than 680 miles an hour.

In review, the Chinese created a dry, lightweight substance that could be packed into confined spaces and yet still ignite with a great deal of force. How did such material come to be employed in weapons? Given the human predilection for warfare, perhaps it would have been a surprise if it had *not* been utilized in combat. Its first use toward this end appears to have had something in common with Greek fire. The Byzantine forces would use air pressure and bronze tubes to squirt this incendiary gunk at an enemy. Some Chinese inventors apparently decided to try and employ their early gunpowders toward similar ends. After all, dry gunpowder was probably easier to store, transport, and handle in general than a flammable liquid would have been. A device known as the firelance was the result of this effort.

Like the Greek fire-thrower, the firelance was based upon the tube design. In fact, the simple tube, eventually transformed into a "barrel," would become a key component of future firearms. In the case of the Chinese weapon, bamboo provided a convenient and readily available source of tubing. One end of the tube was sealed, perhaps with wood, and the hollow cylinder was filled with gunpowder. A wooden staff or rod was attached to the blocked end in order to provide a greater degree of handling and control, hence the name fire*lance*. The rod also further separated the wielder's face and hands from the sparks and heat generated by the device.

In combat, a soldier would light a fuse at the open end of the bamboo tube. This in turn ignited the gunpowder, which would begin to "deflagrate." Although the reaction was fast and violent, the tube played a key role in how the fuel was consumed. Throw a match into a pile of gunpowder sitting on a table, and in a sudden "whoosh" the entire amount will go up in a blinding flash. In the firelance, only one part of the charge was exposed to the ignition, and that limited the amount of gunpowder that could burn at the moment. In addition to controlling the rate at which the gunpowder was consumed, the bamboo played another key role as well. The conflagration that was produced had only one route of escape: down the tube. Sparks and fire therefore shot violently back out of the open end. Theoretically, the Chinese soldier would simply point the blazing stick at his enemies, who would catch on fire.

It is hard to imagine exactly how practical or lethal such an oddity would have been. While the tube did slow down the gunpowder's rate of burn, it still had to have been a relatively short burst of energy. This primitive flamethrower must have had a burn period of no longer than a few minutes. If the enemy temporarily backed out of range, the device could not be turned off and would continue to flare until all the fuel was exhausted. To deal with that very eventuality, many firelances were equipped with spearheads to be used once the gunpowder had burned away. Furthermore, getting the thing to ignite in the first place might also prove problematic, especially in the face of a shower of rain (or arrows). The device does not seem like much of a match against a bow in the hands of a competent archer. One imagines that it might have served better defending walls, towers, or ship gunwales where the effects of enemy missile fire might be better mitigated. Also, in such situations stockpiles of extra firelances could be kept readily at hand. Even in these scenarios, however, it is difficult to imagine how lethal an overgrown sparkler like this would be. It certainly might have inflicted a nasty third-degree burn on exposed skin, but would that halt a determined and well-armored foe? Perhaps the effect was psychological, creating a powerful and semimagical visual display to awe more "primitive" invaders. As we shall see below, certain modifications and additives may help explain the success of the firelance. In fact the device spread far beyond China's borders and enjoyed a particularly long lifespan. A Buddhist painted silk banner from the mid-tenth century appears to show a demon wielding one such weapon. After all, why should a devil settle for a pitchfork when he can use a flamethrower? They were still being employed centuries later against the Turks at the siege of Malta in 1565, in the English Civil War in 1643, and in parts of East Asia as late as the 1930s (Needham 1986, 222–225, 257–258, 261–262).

The rocket also probably evolved from this weapon. Once ignited, the firelance would tend to push back toward its wielder. Someone eventually got the bright idea of turning a lighted lance around, letting go, and sending it sailing off toward the enemy. Nearly 1,000 years later the Chinese used the same principle to send their first astronaut, Yang Liwei, into outer space. Impressive as the rocket is, it may not have been the firelance's only technological descendant. There is good reason to suspect that the incendiary device was also a key ancestor of the firearm.

From a fairly early period the Chinese started to mix projectiles—bits of broken glass and metal—into the gunpowder that was packed in their firelances. Caustic additives such as mercury or arsenic often accompanied the concoction, and when ignited the lance would propel the nightmarish material out toward the enemy. Again, it is hard to imagine glass from an overgrown roman candle taking down an armored enemy, but as noted above the effect of these weapons may have been as much psychological as anything. Deterrence is a curious phenomenon; consider how often people will back away from a violently barking dog a quarter of their size.

The gradual evolution of the gun in this situation can easily be imagined. The firelance's various projectiles were what we call "co-viative" which means that they had a high degree of "windage." In other words, the width of the projectiles did not match the diameter of the tube. Much of the potential force of the propellant therefore slipped around the sides of the objects as they traveled down the bamboo cylinder. The pieces of glass, crockery, stones, fishhooks, fingernail clippings, dried vomit, chicken bones, and God-knows-what-other-nastiness were merely flung at the foe by the force of the fast burn. There really was not much explosive power pushing them.

On occasion perhaps an exceptionally large clump of fish bones or stones might have temporarily plugged up the firelance's tube while it was burning. The slower ignition would have suddenly turned into an explosion, hurtling the projectiles out at a startling velocity. Soldiers might have begun to notice that objects of a diameter similar to that of the tube would tend to get thrown farther than smaller ones. They also might have recognized that round stones or pieces of metal tended to fly farther than flat pieces of irregular glass. Some descriptions of various firelances speak of projectiles being able to kill at several hundred paces. Clearly, in these cases we cannot be talking about co-viative projectiles any longer. Technically speaking, a .69-inch round musket ball does not completely fill an eighteenth-century Brown Bess musket's .75-inch barrel, but the difference is fairly minute. We can presume that some of the medieval Chinese artisans were beginning to approach similar margins of error.

Of course, a side effect of stuffing too much material into a firelance had to have been the occasional accidental explosion. It is entirely possible that something like this is how the explosive attributes of gunpowder were discovered in the very first place. Although the original firelances were most certainly made of bamboo, later models were constructed of wood and metal to reduce the risk of accidental explosion. So was a metal firelance, designed to propel projectiles with little windage, in fact a firearm? It might be very possible to make such a claim.

THE FIRST GUN?

At some point, an unknown person or persons recognized that the potential to cause harm from the projectiles flying out of the end of the firelance was far greater than any accompanying sparks. To capitalize on this effect, our inventors modified the firelance in an important way. Although a fuse run between the projectile and the side of the tube would have been able to ignite the charge, it may have proven a liability by creating a path through which gas could escape around the projectile. Rather than running a fuse into the open mouth of the tube, a "touchhole" was instead drilled into its opposite end. The gunpowder charge could be ignited directly from the rear (or "breech"). Not only would this reduce the opportunity for gas to escape, but it also allowed the soldier more control over the timing of the weapon. A burning fuse run through the mouth of the firearm would have required the shooter to hold the gun in an aiming position, not quite knowing exactly when the device would go off. This would have been more than a bit nerve-wracking. The touchhole solved two problems at once.

So when was this first gun actually built? A curious Buddhist wall carving dated to 1128 appears to show a figure holding a vase-shaped device that is shooting out fire and perhaps a round projectile. The image is not entirely clear, and some skeptics might be inclined to discount it as merely a fanciful carving. On the other hand, we certainly know that the Chinese were engaged in a number of experiments with projectiles and incendiaries. This image may very well represent some kind of primitive firearm. Unfortunately, no actual remains of such devices from the twelfth century have ever been recovered.

The lack of physical evidence might be the result of the material used in the construction of the earliest guns. It is likely that some of the first guns with rear touchholes were still made of bamboo in the manner of a firelance. Even if such primitive guns survived ignition, their biodegradable nature could account for the absence of any surviving artifacts today. Bamboo was also not ideal for the job. There have been various attempts throughout history to create guns out of nonferrous materials, but the high pressures involved have generally rendered these unsuccessful. When we think of guns today, we automatically tend to imagine a metal barrel.

No one can say for certain when the first metal gun was constructed, but we do have an early candidate. The weapon in question was found in Heilungchiang province in Manchuria at the site of a battle that took place in 1288 (Needham 1986, 290, 293–295). If authentic, it was probably used by Chinese troops serving under Mongol commanders. This device is essentially a short bronze tube with a touchhole in the breech. Interestingly enough, the rear of the weapon is rounded and bulblike on the exterior. Internally, the diameter of the tube remains a uniform size. Apparently extra metal was built up around the area where the charge was placed in order to reduce the risk of the weapon bursting. The relatively advanced design of our 1288 "Manchurian candidate" with its reinforced breech implies some degree of experimentation and suggests that earlier, thinner walled models had probably existed too. Another interesting feature of this gun is a socket at the rear of the breech that was almost certainly designed to fit onto a pole. Again, the earliest firearms, being about a foot long and fixed on wooden staffs, seem remarkably like contemporary firelances. The touch-hole and extra bronze at the base appear to be the only major differences. What was the actual purpose of the pole?

In the case of the firelance, the staff was employed to separate the wielder's face and hands from the dangerous material shooting from the mouth of the bamboo barrel. The bamboo likewise might have become hot, and the staff could protect one's hands from being burned. The rod upon which the earliest guns were fastened could have served similar purposes. The staff may also have functioned as an early version of a gunstock. Soldiers might have tucked it under one arm and sighted along the barrel.

It is more likely, however, that the staff was handled in a different manner. In ancient combat the rear end of spears or pikes were often jabbed into the ground, or rested against the side of a foot in the event that the enemy was charging. The idea was for the weapon to remain steady and fixed while the charging soldier or horseman impaled himself on the spearhead. Could the rods attached to early guns have been utilized in the same manner? Given the backward thrust of a firelance, it seems probable that these weapons too were braced against a foot or the earth. Perhaps the first firearms were as well. If the weapon had a heavy gunpowder charge, recoil may have been a serious problem. It seems likely that a tired, hot soldier with sweat-covered hands would have had some trouble keeping his grip on a plain, smooth staff jumping backward at 400 miles an hour. Stabbing the staff into the ground or bracing it with the foot might well have mitigated this problem. It does not seem at all unlikely that early gunners would have simply employed a tried and true method used by their pike-armed colleagues.

There may be historical evidence supporting this hypothesis. In *Bellifortis*, an early European military treatise written in 1405 by Konrad Kyeser, there is an illustration of an early gun being fired. Like the 1288 gun, the weapon is fixed to a wooden pole. The opposite end of the pole appears to be set on the ground, somewhat in the fashion of a pike. The picture is admittedly a touch primitive, and it is difficult to be certain about this interpretation. The weapon in question might also have been somewhat larger and heavier than our 1288 gun, as it is shown being supported by a fixed rest. Furthermore, a later illustration from 1468 seems to show similar weapons being discharged while fully supported in the firers' hands, much like later muskets or rifles. Perhaps a modern-day experiment could shed further light on exactly how these pole-mounted

A sketch of a medieval hand cannon based upon the *Bellifortis* imagery. Drawing by Allison Curry.

guns were fired, provided of course that the field researcher has good health insurance and has his last will and testament in order.

Unlike the telephone or the radio, the firearm's exact birth date is largely unknown. We can estimate, however, that a very basic version of the weapon was on the scene in China by the end of the thirteenth century. The handheld gun merely represented the latest step in an evolutionary process motivated by the desire to inflict damage from a distance. This seemingly universal imperative led humans to develop slings, atlatls, blow-guns, bows, crossbows, and firelances on their path to the firearm. Unlike the mythical Greek goddess Athena, the gun did not emerge fully grown. It was unclear what kind of adult it would grow up to be; there would still be a long and challenging period of development.

2

Childhood: The Rise and Development of Matchlock Firearms

◆

THE HANDCANNON

The importance of the firearm's role in the shaping of the modern world cannot easily be overstated. After its initial appearance in China in the mid-thirteenth century, the firearm underwent a series of significant changes. These were both technological and geographical as the weapon moved out of its birthplace and traveled back and forth across the Eurasian landmass. As a result of some key technical innovations, the weapon's performance increased dramatically and it was adopted by a number of very different cultures. As its usefulness on the battlefield grew, it also came to take an increasingly important position in the plans and strategies of political and military leaders.

One of the most remarkable things about the earliest firearm technology was the speed at which it was transferred to other cultures. Remember that the earliest known firearm dates to 1288 and was found in China. The oldest evidence of a similar weapon in Europe appears in two separate written sources composed in 1326. This technology may have spread right across the Asian landmass within a mere four decades, a rapid trip in comparison to other devices like the crossbow. The Heilungchiang gun comes from Manchuria, which then lay on the far northeastern tip of Asian civilization. One of the two 1326 sources comes from England, on the far northwestern tip of European civilization. One would be hard-pressed to find many more geographically and culturally distinct places in the eastern hemisphere at that time.

How can we account for such rapid movement? Admittedly, firearms were probably around for several decades in China prior to 1288, and some may have been well on their way to the West by the time our soldier dropped his on the Heilungchiang battlefield. Sauce for the goose is sauce for the gander, however, and if we grant several decades of lag-time for the evidence of guns in Asia, we may need to do the same for Europe. It is equally possible that guns were floating around Europe for a few years prior to 1326. There are in fact a few earlier references to European guns, but they are of dubious authenticity.

The thirteenth century was something of a unique time period from a geopolitical standpoint. Temujin, the Mongol leader known as Chingiz (Gengis) Khan, and his successors had carved out the largest contiguous empire in history to that date. At its height the megastate stretched from Korea to the Balkans. Although the empire began to fragment politically in the later part of the century, trade links and cultural connections remained open. During this brief window of opportunity, it may have been fairly easy for the gun to make its way across Central Asia. The Islamic world-trade network probably acted as yet another conduit. It is also unlikely that the transfer of this technology westward was entirely through peaceful mercantile means. There is little reason to doubt that the Mongols were as inclined to use these weapons in western Russia as they were in eastern Manchuria. It is possible that the European and Middle Eastern peoples' first experience with firearms was at the end of the muzzle.

States will go to great ends to obtain the latest military technology. It may be that western cultures were so impressed by firearms that they made extraordinary efforts to buy, steal, or duplicate them. If necessity is the mother of invention, envy may be a parent of imitation. We can surmise that Arabs, Europeans, Persians, Turks, and any others who encountered guns were quick to try and copy them. The current international scramble to obtain nuclear, chemical, and biological weapons of mass destruction (WMDs) may represent a modern example of this phenomenon.

The fact that the term "handcannon" was often applied to the first European guns is somewhat telling. The difference between a cannon and a firearm can be a little bit subjective. In fact, at different points in history the very term "gun" has used been used to designate either weapon. Today we tend to think of a gun as a handheld device, but the origin of the word may be related to the old English term "engyn" as in siege engine. That designation is certainly not appropriate to the firearm, but rather is suggestive of a large weapon designed to hurl heavy projectiles great distances. Cannons more typically fit this image, having been mounted on wheeled carriages or in fixed structures of some type. Smaller or lighter weapons that can be carried into battle more easily are better understood as firearms.

The difference between the two types of guns may exist more in the modern mind than it did in that of the fourteenth-century European. A small bronze cannon unearthed at Loshult in Sweden is believed to be the oldest surviving European gun. Although it cannot be dated precisely, it probably was made sometime in the middle decades of that century. The cannon has a bulbous swelling at the breech designed to withstand the pressure of an exploding powder charge. This shape makes it look much like a mid-nineteenth-century heavy artillery piece such as a Dahlgren gun, built upon similar lines for similar reasons. Unlike those massive cannons of the American Civil War, however, the Loshult gun is quite petite. The diameter of the barrel is 1.4 inches with an overall length of only 11 inches. This hardly matches our modern sense of what a cannon is.

The appearance of the Loshult gun also matches an early European illustration of an apparently heavier, fixed cannon. The picture in question is actually one of the two 1326 sources noted above, and comes from a manuscript by Walter de Milemete written as a guide to his pupil, the future King Edward III of England. In it, an illustration clearly shows an armored soldier igniting the touchhole of a vase-shaped weapon. It looks somewhat like the Loshult gun, although it is much more rounded and bulbous and apparently somewhat larger. In fact, the weapon it seems to have the most in common with is that depicted in the 1128 wall carving! Unlike the handheld Buddhist cannon, the Milemete weapon appears to be resting on some kind of table-like structure or simple gun carriage. Using our working definition above, the picture seems more suggestive of a cannon rather than a handheld firearm. Incidentally, the other 1326 source is a decree from Florence appointing two men to make "cannons of metal" (Reid 1984, 43–45).

Of further interest is the projectile coming out from the barrel of the weapon in the Milemete illustration. Rather than a cannonball, it is an enormous arrow. The Florentine decree also requested iron arrows, probably as ammunition for the cannons being ordered. The fact that early European cannons fired metal arrows implies the dependency of new weapons technologies upon earlier, previous systems. A society in which long-range warfare was dominated by bows and crossbows would naturally turn to familiar forms when fashioning new armaments.

There are a number of additional references to cannons in the 1330s and 1340s. Most are linked to some kind of attack on a castle, town, or fixed fortification. However, the English are believed to have used guns at the Battle of Crécy in 1346 and at Poitiers ten years later. Both of these en-counters were field battles between mobile forces and did not involve a long siege. It is curious that at this early point in their development, some guns were being used in a fairly mobile fashion.

Back in China, early guns and cannons continued to be constructed. Another barrel found there dates to 1332. If there are some doubts about the 1128 wall carving and the 1288 gun, little exist regarding this later weapon. In this case, the date of manufacture was actually inscribed directly upon the barrel. It might be noted that in the past, evidence of Asian gun technology was often subjected to a much higher degree of scrutiny and skepticism than was that of the Europeans. For centuries, Westerners cred-ited an imaginary medieval monk known as "Black Berthold" with the in-vention of both gunpowder and guns. Fortunately, solid historical evidence from China leaves little maneuver room today for such nonsense. The 1332 model is very small, like the earlier Manchurian gun, but its basic design is somewhat different. Instead of sporting a bulbous powder chamber, the shape of the barrel is a straight, even cylinder. Later Chinese designs follow this model as well.

Interestingly enough, the straight design likewise began to appear in Europe too. Most late-fourteenth- and early-fifteenth-century guns had a more uniform and less "swollen" breech. They compensated for the pres-sure of the explosion at the rear with a separate powder chamber. This chamber was smaller than the diameter of the barrel, which in effect left more metal around the powder charge itself. Some of these weapons, al-though not bulbous, still remained slightly thicker in the breech to com-pensate for the higher stress.

The oldest surviving European gun that has been definitively dated matches this later design. It was found in the ruins of Tannenberg Castle in Germany, which was destroyed in 1399, so scholars can safely presume that it was manufactured prior to this time. It had a slightly larger dimension at the breech than in the muzzle of the barrel and used a reduced powder chamber. Some models made only a few decades later are completely uniform in the exterior of the barrel. If we start with the gun illustrated in the 1326 Milemete manuscript, then consider the Loshult gun, the Tannenberg gun, and later weapons of the fifteenth century, there is a clear movement from the rounded, bulbous shape to the more cylindrical barrel we are familiar with.

One must be careful to avoid generalizing too much about these guns because of the great variety of shapes and designs that emerged. There was little in the way of standardization. One early European gun, also recovered in Sweden, even has a small bronze bust of a bearded man attached to the top of the barrel. This may have represented the gun's owner, a saint, or even Christ. Perhaps the idea was to let the enemy get a glimpse of Jesus before going to meet him in person. Some of the early guns, like the Tannenberg weapon, continued to have a rear socket so they could be attached to a staff, sometimes called a "tiller" like the crossbow's main component. Others were designed to be strapped down horizontally to the end of a primitive wooden stock. Some were even completely metal and had forged staffs tapering out from the breech of the barrel. Many were bronze, but iron handcannons became increasingly popular, probably due to their lower cost. Early barrels were initially cast as a single piece, whatever the metal used. A final common characteristic of these first European firearms was a hooklike extension protruding from the bottom of the barrel. Apparently the idea was to fix the hook over a rampart or gunwale when firing. This would help stabilize the piece, and may have also reduced its recoil.

The split between the firearm and the cannon started to become more apparent late in the fourteenth and early fifteenth centuries. Initially it seems as if guns were harassment weapons designed to throw metal arrows and other projectiles over defensive walls in order to demoralize the enemy inside. Gradually it was recognized that spherical shot could actually inflict damage upon castle walls, town fortifications, and ships. Heavier and heavier versions were constructed to this end, and by the mid-fifteenth century truly gargantuan devices had been built that were capable of firing stones weighing 1,000 pounds (Keen 1999, 273–274).

It is clear in the emergence of such massive artillery why terms like "hand gun" or "hand cannon" came into use. The function of heavy siege weapons is pretty clear, but what role did their smaller cousins play? They were certainly easier to carry, but it is difficult to imagine that they were particularly dangerous. Like the firelance, perhaps the early handcannon's function was largely psychological. It seems unlikely that it was more lethal than longbows and crossbows, and it was unquestionably less accurate. Furthermore, the gun had to have been tricky to aim, particularly given the fact that this was done with one hand. The other hand was used to ignite the charge, probably with a bit of burning wood, cloth, or, more likely, a glowing hot wire. This required a small nearby fire, which clearly would have limited the weapons' flexibility. The low effectiveness of the handgun is best proven by the continued popularity of bows. After all, the famous English victory at Agincourt in 1415 was dependant upon the longbow, not the handcannon. In that encounter thousands of combatants were brought down by arrows, but only one Englishman was shot dead by a "gunstone" (Hardy 1992, 114).

THE LATE MEDIEVAL FIREARM

Nevertheless, throughout the remainder of the fifteenth century longbows went into gradual decline. Crossbows remained popular, but increasingly firearms came to be viewed as a primary projectile weapon for infantry. At some point in the fourteenth century, firearms apparently began to actually kill people with some regularity, but there is more to the story than just that. Apparently a number of factors were at play.

First of all, a new technique of manufacturing gunpowder was developed around this time. This process, known as "corning," greatly strengthened the power and performance of gunpowder. Early gunpowders should have come with a consumer warning: "some settling of contents may occur." Over time the heavier nitrates in gunpowder tended to settle toward the bottom of the mixture while lighter-weight charcoal stayed on the surface, which greatly affected its performance and shelf life.

Corning is the process of turning light, dusty gunpowder into a granular form. Gunpowder was wetted, sometimes with alcohol, and rolled out on sheets to dry. It was then broken up into a number of kernels, granules, or grains. These could be sifted, filtered, and sorted into their different grain sizes if desired.[1] It was gradually recognized that different kernels or grains had distinct advantages or disadvantages depending upon the application. More importantly, each small granule contained the right proportion of nitrate, sulphur, and carbon. Corned powder was much less prone to breaking down into its constitute elements. Furthermore, the small irregular grains allowed more air to filter into the mixture, further promoting its flammability. The new granular form of gunpowder was significantly more explosive and powerful than the earlier, loose powder. Some contemporaries estimated that it had three times the explosive potential. Later on, Europeans would even figure out how to apply a coating of graphite to the mixture, which made it much more resistant to moisture (Hogg 1996, 21–22). Corned gunpowder was more dependable and lethal; a bullet propelled by this stuff could punch through most armor.

Guns might have also become more popular for some less than logical reasons. Contrary to the popular image of science and technology, change and innovation are not always based upon pure models of efficiency. Like so many elements of human life, technology is affected by society and social culture. Consider the decline of the typewriter in recent years. In

---

1. The term "grain" is also used as a measurement of weight, with 1/7000th of a pound equaling 1 grain. The two terms can sometimes become confused when discussing the number of actual kernels in a measure of gunpowder versus its weight in grains.

most ways the computer is a vastly superior technology, but it must be recognized that typewriters do not crash, lock up, or delete papers. Still, millions of people racing to embrace the latest thing have been willing to overlook the personal computer's obvious shortcomings. The widespread adoption of firearms may also be another example of this phenomenon at work.

Guns were clearly showing their potential. The French had lost most of the major field battles of the Hundred Years' War, but in the 1440s they began to concentrate on the use of artillery to knock down enemy castles and fortifications. Eventually these weapons would drive the English off the continent. Cannons were new, powerful, loud, and flashy, so who would not want a smaller version? People probably associated handcannons with their larger cousins and with being part of the very latest developments in warfare. Certainly the bang and the smoke of firearms had a psychological effect on the enemy and gave an adrenaline rush to the firer.

In any event these weapons became increasingly common and effective on the battlefield. They were used to particularly good effect in the wars of the Hussite religious reform movement from 1420 to 1434. The Hussites' blind yet farsighted Czech leader, Jan Zizka, created a mobile army of fortified wagons. When attacked, they circled up like an American settler wagon train or a Boer kraal, and fired out at the enemy using a combination of crossbows, handcannons, and small artillery. These unorthodox methods completely confounded the German crusader knights vainly trying to crush the heretical movement. The Florentine historian Pierino Belli also described the effective use of handcannons in the siege of Lucca in 1430. The weapons apparently made short work of any armor or shields they encountered, and some lucky shots supposedly pierced through as many as three men (Brown 1980, 13–14). At Caravaggio in 1448 the smoke from Milanese handgunners reportedly was so heavy that it obscured the battlefield, a commentary that implies the presence of many firearms.

The increasing success of early gunsmiths promoted a response amongst armor makers, and heavy armor came to have a symbiotic relationship with firearms for a while in the late Middle Ages and the Renaissance. When we think of medieval times, popular images of the knight inevitably include very heavy and extensive armor. Nevertheless, full-plate mail really did not appear until the late medieval period, in response to the rising power of missile weapons. As handcannons and later firearms became more effective at breaching armor, stronger and more elaborate suits were devised. As this metallic apparel became stronger, firearms likewise became more and more efficient and effective. Some suits of armor even came to include a "proof-mark," which was a small dent where a test bullet had been fired into the

A seventeenth-century Dutch gunsmith in his workshop. © Hulton Archive/Getty Images.

chest plate. Ultimately, this was a losing battle as it was generally easier to develop more powerful handguns than it was to forge elaborate metal clothing.

By the early fifteenth century, these handguns had become weapons that were clearly recognizable as firearms. They used the explosion of gunpowder to fire projectiles. Each one had a metal barrel and a very simple version of a stock. A few simple changes next occurred that helped the weapons evolve into even more familiar forms. The handcannon design with the barrel strapped horizontally down to the wood tiller won out over other models with a staff attached to a socket in the rear of the barrel. One might suspect the earlier version was more prone to breaking. The tiller itself also began to assume a more familiar gunstock shape. The stock also became somewhat shorter and thicker, and was clearly intended to be braced in the arms when fired. The barrel of the gun was fitted into a groove carved into the stock, and the tubes themselves also grew longer. A cannon-firearm hybrid called the "culverin" discharged a ball that was only .70–.80 inches in diameter but had a barrel between 5 and 6 feet long. It was pretty heavy, and required a crew of two to operate.

This covers the rise of more modern stocks and barrels, but not locks. How did the lock move from the crossbow to the gun? It was linked to ignition. Hot wire was soon replaced with a piece of rope or cord pretreated with a saltpeter solution. This cord burned slowly and steadily when lit and became known as matchcord, or "match" for short. It was these smoldering cords that the pirate Blackbeard allegedly tied into his hair, *not* household kitchen matches. Just as William Teach could run around with a smoking head, the typical gunner could carry a burning strand of this stuff around without remaining near a fire. This may not seem like a big deal today, but it led to further developments in the gunlock.

The next logical step would be a device to remove the match from the gunners' hands entirely while shooting. The end product was known as the serpentine or serpentine arm, an S-shaped piece of metal fastened to the side of the stock near the breech. One end of the S extended below the gun while the other reached above it. The upper end contained a clip, which held the burning tip of the match in place slightly over and behind the touchhole. When the firer pulled back on the lower end of the serpentine, the upper end brought the burning match into contact with the touchhole and the gun discharged. This allowed gunners the luxury of being able to more carefully aim their weapons along the barrel to line up a shot. Some of the serpentine firearms of the mid-fifteenth century actually had gun sights for aiming. These consisted of a small front blade mounted near the muzzle of the gun, with a second notched block at the rear. The forward sight was aligned with the notch and pointed at the enemy before the weapon was fired. Some have speculated that this aiming sight was based on an older technique used to sight early crossbows. The thin blade may have represented the arrow point, while the notch was supposed to imitate the indentation of the trigger hand between the firer's thumb and first finger. This same basic system is still used for many open-sighted firearms today.

THE MATCHLOCK

One might say that by the late fifteenth or early sixteenth century, the firearm was beginning to approach a sort of adolescence. Questions of identity can make this a confusing stage in a young person's life and, allegorically speaking, so too was it for the firearm. Throughout the fifteenth century, there was a proliferation of different types of guns. A variety of terms were employed in various parts of Europe to describe an equally diverse number of weapons. Sometimes the name referred to a weapon's size, its function, or its firing mechanism. For example, the term "musket" originally designated a very heavy firearm, whereas "wheel lock" was a type of firing mechanism. One might have a wheel lock fixed to a musket, but apply either (or both) names to the same weapon. To make sense of all this, we will simply try to stick to the most commonly accepted modern designations for these firearms, such as "arquebus."

That warning aside, the next step in the development of the firearm was something commonly called the matchlock. As might be expected, disagreements exist between modern writers over what constitutes a matchlock and what does not. Some authors have argued that the serpentine is a type of matchlock, while others have advocated for a much narrower definition.[2] Technically speaking, neither the serpentine nor the earliest matchlock weapons had any actual *locking* mechanism on them per se, but they did have a firing apparatus that accomplished a similar task to that device found on a crossbow.

The most common version contained a refashioned serpentine arm that looked like a reversed "C," with the open part of the letter facing the shooter. On the side of the stock near the breech of the barrel was a flat piece of metal called a lockplate. The upper end of the serpentine arm held a smoldering match, while the lower extremity was attached to the lock mechanism hidden behind the plate. The system was a little different than that found in a crossbow. A metal leaf-spring normally held the serpentine up in place, but when the large barlike trigger was squeezed, the serpentine correspondingly moved downward toward a small basinlike structure. This was the aptly named "flashpan," which was fitted to the outside of the touchhole and filled with a small amount of gunpowder. If everything went right, the ignition in the flashpan (or "pan") would travel through the touchhole, then down into the barrel, and in turn set off the main powder charge. The origin of the old expression "just a flash in the

2. For example, William Reid (1984, 51) seems to make no distinctions between serpentines and matchlocks, while M. L. Brown (1980, 24–26) is quite particular in his definitions.

pan" to describe a disappointing or unfulfilled event arose from cases in which things did not go quite this way. It is also obvious how, in this early design, the mechanism really did not "lock" prior to shooting and squeezing the trigger did not trip any kind of sudden motion. Later matchlocks were often fitted with a retractable flashpan cover intended to minimize the effects of bad weather. Some models also included a flat piece of shielding called a "fireguard," which was designed to protect the shooter's face from the ignition's conflagration.

Just as the ignition mechanism of these weapons changed, their gunstocks continued to grow longer and more elaborate along with the firearm. They could have a fairly straight tillerlike design, or much more of a curved and hooked stock. Both curved and straight stocks were flattened at the far end with an eye toward bracing the gun. The butt was set against either the shoulder or the flat of the chest. In the latter case, we can imagine that the recoil of one of these firearms against the sternum could not have been a pleasant experience. On the other hand, some of these early guns were fairly heavy and the weight of the weapon would have helped absorb this "kick."

Much of such weight, of course, came from the barrel. While locks and stocks were changing, the barrel of the common gun remained a basic tube. Instead of producing gun barrels by single-piece casting, a different technique known as "hammer welding" or "forging" had evolved. Through this method, a flat, narrow strip of iron about a foot and a half in length was carefully hammered around a steel rod known as a "mandrel." The two lengthwise sides were welded together to form a tube, and the mandrel was then removed. In order to visualize this, imagine connecting the open sides of a hot dog's bun together and slipping the meat itself out of one end. Two or three of the resulting iron tubes were next welded together end-to-end to form the complete barrel. The barrel then sported a longitudinal seam down its length that would be filed down and polished. The tube could be further bored out to a desired diameter and the breech sealed by welding or with an iron plug. A more difficult technique involved drilling the bore out of a solid cylindrical bar of iron, but this method was not popular due to the extensive effort involved. It would become common centuries later when industrialization made this process more automated and efficient. In the meanwhile, forging was an effective enough method to provide barrels of different lengths and diameters for a variety of guns.

Remember that there was little standardization to these firearms, but some common terms came to be employed to describe matchlock weapons of similar size and function. One of the more popular matchlock firearms was a weapon called the arquebus. Although there were a number of variants to this name such as "harquebuse" or "hackbut," the appellation most likely was a French derivative of the German phrase "hakenbuchse," meaning "hook gun." "Buchse" was shortened to "bus," while "haken" was replaced with "arque," a French term for arch or hook. Like earlier handcannons, many arquebuses continued to carry such protrusions below the barrel, and they may have been the source of the appellation. Likewise, the hooklike appearance of the serpentine may also have contributed to the term. Finally, some arquebuses were fitted with a curved or hooklike gun butt as noted above, which may have been a third inspiration for the name.

The term "arquebus" was a fairly standard appellation for most handheld, lightweight guns, but 1499 saw the first recorded use of the term "moschetto" in Italy to describe a particularly heavy weapon. This firearm was named after a bird, specifically, a male sparrow hawk. In England, the term was transformed into the more familiar "musket" (Reid 1984, 64). The fifteenth-century musket was an exceptionally heavy gun, the descendant of the culverin. It could weigh 20 pounds and sport a 4.5-foot barrel. Such monsters were not handled easily, and they required some kind of support to be properly aimed. This might take the form of a crude tripod, but typically a rod with a U-shaped fork at one end was used. The barrel of the gun rested in the fork, while the other end was driven into the ground. Supports like this were not new and were employed in some of the earliest handcannons, such as the *Bellefortis* image noted in the previous chapter. In the early fifteenth century, the fork was sometimes held in place by a small residual hook protruding from below the barrel, although this disappeared in later models and the supporting stick was simply secured by the firer's hand. One other accessory found on muskets and most other early modern firearms was the ramrod, used to push the bullet down the barrel into the breech. This thin, rodlike device was typically made of wood and could be conveniently stored in a channel or groove running along the bottom of the stock directly below the barrel.

Another common fifteenth-century firearm was something called the "callivern" or "caliver." This was a midranged weapon with a size between that of the arquebus and the musket. It did not use a forked rest, and it fired a slightly smaller caliber projectile than did the musket. In fact, while the name "caliver" may have been linked to the earlier culverin, both words might have been corruptions of the word "caliber" itself. The latter term, used today to describe the diameter of a barrel or a projectile, may have evolved from the Latin *qua libra* or the Arabic *q'alib*. The first term is a reference to weight and could have been linked to

A sixteenth-century Dutch arquebusier. The hooklike shape of his firearm's stock is clearly evident. © North Wind Picture Archives.

projectile size, while the second means "mold" and perhaps was connected to a bullet mold. Caliber has traditionally been expressed as a percentage of an inch; thus, a gun with a diameter of half an inch is referred to as a .50 caliber. Today, caliber is also often expressed in metric millimeter units, and a .50-caliber bullet might likewise be called a 12.7mm. Besides the arquebus, the musket, and the hybrid caliver, there was a variety of strange firearms, some of which defy easy categorization, such as an odd combination of a shield and a gun found in the arsenal of Henry VIII of England.

Most of these weapons, utilizing corned powder and matchlocks, could be quite lethal in battle. No longer was the effect merely psychological. While they could all punch through armor, the massive musket reportedly could do so easily at well over 100 yards. The murderous device was also said to be lethal to an unarmored man or horse as far away as twice that distance. The effective reach of any single weapon would have been nowhere near this range, but if a large number of musketeers were firing into a sizeable group of the enemy, somebody was bound to get hurt. These early firearms also inflicted wounds that were generally more savage than those caused by arrows or hand-to-hand weapons. In particular the bullets, made of soft lead, tended to shatter bones into small fragments. Modern orthopedic surgeons would be hard-pressed today to repair the effects of a direct hit; in that earlier time period, the usual course of treatment was amputation.

The power of these devices helps explain the growing popularity of firearms in the late fifteenth and early sixteenth centuries. In 1471 the army of Charles the Rash of Burgundy contained one handgunner for every four archers, a statistic some authors cite as evidence of the persistence of bow weapons (Parker 1988, 17). This is something of a "glass half-empty" argument, and the data can be interpreted in a different way; after all, one in four is certainly a higher percentage of guns than were carried by earlier armies. One must admittedly recognize that this was a gradual change, however, not an overnight transformation. For example, Charles's missile weapon–oriented army was crushed in several disastrous encounters with Swiss pikemen, a fact apparently at odds with the growing trend toward firearms.

The trend did continue nonetheless. By 1490 Venice started to replace all their crossbowmen with handgunners. Several years later Gonsalvo de Cordoba, a commander in the service of the Iberian monarchs Ferdinand and Isabella, began to reorganize his forces with more attention to firearms. This was in response to serious losses inflicted by Swiss pikemen at Seminara in 1495. The payoff came eight years later at Cerignola when the Swiss, in the service of the French crown, launched one of their classic frontal assaults. Gonsalvo's troops had cut a ditch in front of their lines and threw up fortifications prior to the attack. His arquebusiers dramatically repelled the onslaught in a battle that may represent the first real victory attributable to the firearm. The lesson was not fully learned, however, and several more massive clashes pitting pikemen against handguns and cannons were fought. The most disastrous was at Bicocca in 1522, where some 3,000 Swiss went down while assaulting Spanish positions. Three years later arquebus-wielding troops were used again to good effect at Pavia. In this mobile battle, they did not rest behind field works but operated effectively as maneuvering troops out in the open, proving that the firearm was a flexible weapon. The pike was not completely abandoned in the face of these changes, however. Pike-equipped units continued to be useful in hand-to-hand combat, but the ranged fighting was now primarily conducted by firearms.

By the early sixteenth century, the mounted knight had likewise fallen from his once lofty position dominating the battlefield. The shrewd Renaissance political theorist Niccòlo Machiavelli seemed well aware of this reality. In his *Art of War*, written in 1520, Machiavelli remained a bit skeptical about firearms but thought even less of horsemen. He noted that people or kingdoms that valued the cavalry more than the infantry were weaker and more exposed to complete ruin (Machiavelli 1520). His attitude toward mounted troops reflected the fact that this was an admittedly bleak period for knights, and they knew it. One knightly commander, Gian Paolo Vitelli, felt the solution to the technological challenge posed by the firearm was to simply remove the hands and eyes of captured handgunners. Beyond merely illustrating how psychotic Vitelli was, his prescription suggests a degree of desperation in the face of changes he could not control. A similar illustration of this response is found in the character of Pierre du Terrail, chevalier de Bayard, who was said to be "without fear and without blame" (Keegan 1994, 333). This grand knight once supposedly held a bridge single-handedly against 200 enemy soldiers, but did not extend the concept of chivalry to captured crossbowmen or handgunners. He had these men executed, sometimes by shooting in order that the punishment might better fit the crime. Occasionally fate does have a way of catching up with people, and in 1524 he was shot dead by a lowly musketeer. While this particular cavalier disappeared with a bang and a flash of smoke, the mounted warrior in general was not quite finished. Their armor would shrink and their lances were generally replaced by sabers, but elite troops on horseback would remain around for centuries. Surprisingly enough, another version of the firearm would have at least some role in the survival of cavalry as a branch of military service. More on this will be reviewed in Chapter 3. Nevertheless, as a central force on the battlefield, the medieval knight had been measured, weighed, and was most definitely found wanting.

THE INTERNATIONAL PROLIFERATION OF THE MATCHLOCK

By the early sixteenth century, guns had already undergone great technical changes. The firearm may have been born in Asia, but it was on the battlefields of Renaissance Europe that it really came of age. The use of this particular gun was not restricted to western forces by any means, however, and the matchlock quickly migrated back through Asia. It is also possible that Asian peoples independently developed their own versions of this gun. The device did seem to appear suddenly in Middle Eastern arsenals at about the same time it was first being used in the West. The first serpentine versions of the matchlock may have owed their development to Turkish gunsmiths (Pacey 1990, 74–75). Future discoveries or investigations hopefully will further clarify the picture. In any event, the matchlock had little trouble finding employment across the Asian continent in the sixteenth century.

As a result of this diffusion, a number of powerful states emerged that based their success upon matchlock firearms and cannon. Historians have thus come to describe these entities as "gunpowder empires" (McNeil 1982, 95–99). While artillery clearly was a key formula for their victories, the matchlock played a critical role too. Once again, the widespread appearance of these weapons outside of Europe belies the idea that the use of firearms was an exclusively western phenomenon. In fact, European peoples were a common target for at least one of these powerful states.

The Ottoman Turks were well aware of gunpowder's power. Their siege artillery enabled them to take down the second Roman capital, Constantinople, after it had defied similar attempts for over 1,000 years. Constantinople still had an imperial destiny, however, and after 1453 it became the heart of a new, vigorous, Islamic empire. This state was highly mobilized toward the pursuit of war. In addition to the widespread use of cannons, Ottoman sultans were regularly equipping their troops with matchlocks by 1500. These firearms were successfully utilized in campaigns against Persian horseback-mounted forces in 1514, and against the Mamluks of Egypt a few years later. A famous exchange occurred in the second campaign when a captured Mamluk leader, Amir Kurtbay, taunted his Turkish counterpart, "A single one of us can defeat your whole army.

If you do not believe it, you may try, only please order your army to stop shooting with firearms" (Ayalon 1956, 94). The Mamluks felt that the Ottomans did not fight fairly with swords and lances, but somehow cheated by using guns. The Egyptian rulers held these devices in such contempt that they only issued them to their lowest ranking slaves.

Turkish matchlocks were similar to those found in Europe, but in the early sixteenth century their gunsmiths did develop a distinctly different manner of constructing barrels. They used the technique of winding or coiling very long thin strips of red-hot steel around the mandrel at a slight angle. These strips were then fused together, and the resulting barrel was thus constructed somewhat like the modern cardboard tube found inside a roll of paper towels. After forging, the barrel was treated with light acid, often citrus juice, to highlight the decorative spiral pattern. These barrels were less prone to bursting than the longitudinal forged barrels of Europe. The Ottomans also had access to exceptionally high-quality steel from India and Iran, making their barrels even stronger. The barrels became associated with the city of Damascus in particular and eventually became known in Europe as damascus-twist barrels.

As noted above, the idea of dangerous weapons technology being passed around the globe is a major concern today. Such transfer of military technology is not new, however. European craftsmen presumably were an important source of Turkish weapon technology. For example, a key technical advisor during the siege of Constantinople was said to be a cannon-maker known as Urban of Hungary. Unhappy with the pay offered by Christian leaders, he found better employment with the sultan. Likewise, Ottoman bronze casters and gunsmiths had occasion to travel abroad and spread their knowledge too. One of these, Husain Khan, reportedly helped spread weapons technology throughout the Islamic world, even to distant Sumatra (Crosby 2002, 116–117).

Much of the Ottoman skill with firearms and cannons was picked up in turn by the Mughals, a people whose leaders claimed descent from the Mongols but had come to embrace the religion of Islam and the culture of Persia. Mughal commanders chose to base their forces not upon a Mongol-style cavalry army, but rather adopted Ottoman tactics, cannons, and matchlock firearms. Using this technology, they were able to carve out an empire on the Indian subcontinent. In 1526 and 1527 the first Mughal emperor, Babur, won key victories at Panipat and Kanua against the indigenous Indian princes or "rajas." These monarchs, who employed tens of thousands of horseback-mounted troops and hundreds of elephants, were no more successful against firearms than had been Egyptian Mamluks or European knights.

The Mughals were just one of a number of early modern peoples linked by heritage to the earlier Mongols. Another branch of the family tree known as the "Golden Horde" still dominated much of northwest Asia and parts of Europe. One of their tributary states was the Principality of Muscovy, which was led by Ivan III in the late fifteenth century. Using his trade contacts with Europe, Ivan began to import large numbers of matchlock firearms. In what was becoming an often-repeated story, the horse-mounted warriors of the Golden Horde were unable to prevent Ivan from driving them off their western holdings and declaring himself ruler off all the Russias. The expansion of Moscow continued under his son Ivan IV, known as "the Terrible." His firearm-equipped armies continued to rage eastward against their former masters. By the time this new tsar, or "caesar," died in the year 1584, Russian adventurers had carried the matchlock across the Urals into the Asian frontier on a steady drive toward China and the Pacific.

How would the Chinese handle the Russians when they eventually began showing up on their northern border a century or so later? Well, one might expect the homeland of the firearm to have produced a number of more advanced and innovative designs by the sixteenth century. In fact, it appears that this may not have been the case. Perhaps the relative stability and unity of China had not encouraged the gun's further development. Jesuit missionaries operating in China in the late sixteenth and early seventeenth centuries found the Chinese to be highly interested in European cannons and firearms that were now recognized as superior. Widespread reorganization of Chinese forces upon the lines of a gunpowder empire did not really occur until the conquest of China in 1644 by barbarian outsiders, the Manchus. Perhaps because these northerners were initially less tied to Chinese traditions, they made more frequent use of firearms, primarily matchlocks. This helped Manchu-ruled China to expand eastward into Tibet, Xianjiang, and Mongolia. The weapons also helped to neutralize the Russian threat to the north and eventually led to a treaty stabilizing the border between the two powers in 1727.

One of the most curious episodes in the history of firearms developed further to the east of China. The Japanese had trade and cultural contacts with the larger state since time immemorial, and were undoubtedly familiar with gunpowder. Around 1543, however, they began to encounter European firearms for the first time. Portuguese traders and their arquebuses created a great deal of interest. Japanese swordsmiths were among the finest in the world at that time, so there was some degree of readily available metallurgical skill to be tapped. Copying the foreign matchlocks was a relatively simple matter. The islanders soon had their own domestic firearms industry, and they chose to modify the arquebus's design in a few ways.

Generally, the Japanese version was of a larger caliber than the European model. They also designed a small hood for the end of the serpentine to protect the burning match from the elements. The firing mechanism would be altered as well, employing a reliable system that did actually "lock" the serpentine upright to be released when the trigger was pulled. Another unique feature was the placement of the lock mechanism itself. In Japanese models, the mainspring and other lock parts were fixed on the *outside* of the lockplate, perhaps to make repairs or cleaning easier. On the downside, the components were often made of inexpensive brass that did not hold up as well as steel. Such a lock would require more frequent servicing, which perhaps explains why they used an external structure. Some critics also found the Japanese touchhole positioned too close to the shooter's eyes (Pacey 1990, 89).

The reason why feudal Japan was so interested in these new weapons had much to do with domestic politics. The islands were politically fractured at that time, and local feudal lords, known as "daimyos," were interested in gaining any advantage they could over their rivals. One noble who exploited the arquebus to its fullest extent was Oda Nobunaga. He began equipping his troops with them in 1549 and encouraged the further domestic production of matchlocks. He also took steps to promote trade with the Portuguese in order to secure dependable supplies of lead and saltpeter. This effort ultimately paid off in 1575 at the battle of Nagashino, where 3,000 of Nobunaga's arqubusiers won a decisive victory. Interestingly enough, they deployed in three ranks and fired in succession, a technique Europeans would later develop too.

Through the efforts of Nobunaga and two other important daimyos, Toyotomi Hideyoshi and Tokugawa Ieyasu, Japan was unified by the early sixteenth century. Now that the political situation had changed, the nation's new leaders made an effort to disarm the country in order to prevent further chaos or potential rebellion. They called for guns (and other weapons) to be melted down into an immense statue of the Buddha. They also promoted the old samurai cult of the sword and restricted the manufacture of matchlocks. This policy was so successful that Japan became one of the few nations in history to effectively and intentionally turn its back upon the firearm.[3]

We clearly see in the case of Japan how inventions can be affected by political and social developments. The way a technological system emerges

---

3. For more information on this, see Perrin (1980).

and grows is obviously the end result of human imperatives. Recognizing this helps to explain where, how, and why the firearm changed in the way that it did. It was sought by those who wanted a weapon that required relatively little training, but could punch through enemy armor at a distance. Technical evolution does not always follow a perfectly logical path, but then neither do human beings. Culture and politics may help us to understand why one Asian nation was discarding the firearm, while at the same time European peoples were continuing to "improve" it.

3

Between a Rock and a Hard Place: The Rise of the Flintlock

◆

MATCHLOCK EQUILIBRIUM

The last chapter reviewed how simple handcannons evolved into much more familiar matchlock weapons that incorporated all three key components of firearms; locks, stocks, and barrels. By the end of the sixteenth century, the basic matchlock arquebus could be found throughout the breadth and length of the Eurasian landmass. There would be little stability in the firearm's life cycle, however, as Europeans continued to experiment with new ignition mechanisms for the weapon.

Most people today would admit that we live in an age of disparity. Scholars, journalists, and other really smart people focus upon the gap between nations with so-called advanced economies and those countries whose economic structures are supposedly in the process of "developing." The contention that there are uneven levels of financial, social, or technological success among peoples is not particularly new. One area of human activity that has shown great inequality among societies is the conduct of warfare. In recent years, two regimes in the Middle East have been quickly and decisively driven from power largely through the use of high-tech weaponry. In those cases where their soldiers were brave enough or coerced enough to attempt a field battle, the results were largely one-sided. This disparity in conventional arms has led some forces of opposition to adopt radical and subversive terrorist tactics against western nations. A global imbalance of military power, however, has not always existed. As late as Agincourt in 1415, one of the most lethal weapons in Europe's arsenal was still the ubiquitous bow and arrow. Given this weapon's worldwide distribution, it seems logical to assume that a reasonably general level of international parity in military technology existed at that time. This technological parity began to come apart in the sixteenth century with the advent of matchlock firearms. Those societies that adopted the arquebus (and cannons) now began to enjoy a fairly serious level of military supremacy. This fact was driven home to the Golden Horde and to the soldiers of the raja of Dehli, who had to face Russian and Mughal guns.

Still, by the discrepancies of today the international military imbalance found in the sixteenth century does not seem quite so extreme. Yes, there were several new empires carved out at least in part by firearms. But the key word here is "several." Rather than one or two superpowers, there was a handful of states equipped with essentially the same military technology. Perhaps even more significant is the distribution of these powers. We have already reviewed three major gunpowder empires of the 1500s: those of the Mughals, the Ottomans, and the Russians. The empire of the Spanish and Austrian Habsburgs should probably be included as a fourth. In this schema, half of the major gunpowder empires were based outside of Europe. Admittedly there were other nonimperial states in Europe that possessed matchlocks, but certainly the case can also be made that Japan was a similar entity.

The purpose of this survey is to point out that there was no real European monopoly on advanced firearm technology at least through the sixteenth century. This became less and less the case as time progressed. Some minor distinctions in the firearms emerged fairly early on. Of course, the stronger damascus-twist barrels of the Ottomans were widely prized, but European lock mechanisms were also considered universally superior. This created a good opportunity for trade, and often Turkish and Indian guns came to be equipped with western ignition systems, while the finest Italian or German firearms supported twisted barrels of Turkish design.

Firearms in Asia admittedly continued to evolve on their own. In India, for example, a number of unique matchlock designs emerged, some with a mostly internal serpentine. While the style of gun could vary from region to region, the basic technique of ignition by match remained the same. In the homeland of the firearm, things were even more static, and Chinese matchlocks were largely imported (or copied) Japanese weapons. For their part, the guns of Japan changed relatively little from the sixteenth century until the arrival of well-armed American thugs led by Matthew Perry in 1853.

Those Yankees and their weapons were the heirs of a western technological legacy that had developed much more lethal designs. The equilibrium that kept the matchlock-equipped arquebus predominant in Asia did not exist in Europe, where new firearm technologies emerged. This discrepancy is really part of a much larger and broader question about European hegemony over much of the world in the past several centuries. How did this backward, fractured, and peripheral peninsula come to dominate so much of the globe by the twentieth century?

In the past, Europeans themselves pointed to various issues such as religion, culture, race, or economics as the secret of their success. Other historians have stressed technology itself. While that angle has some appeal in a technography like this, it creates something of a chicken-and-egg issue. Was superior technology the cause or the result of western success? In other words, why did Europeans embrace this Chinese invention so heartily? What led them to desperately want to improve upon it? The firearm itself certainly did not demand to be perfected. Some persons, some human beings, had to wish it so.

Perhaps the best answer lies in the geopolitical situation that Europe found itself in. China was undergoing a wave of barbarian (Khitan, Jurchen, and Mongol) attacks at the time they were experimenting with firelances and guns. Once the outside aggression had subsided, however, China saw little reason to keep experimenting with improved firearms. Japan is an even better example of this phenomenon. Once matchlocks had completed their job of unifying the country, the samurai had few motives to promote their use.

Europeans, on the other hand, were in a different situation. Unlike China, Europe had no unified culture or common political ruler. Their enemies were not barbarians at an imaginary continental gate but rather were ambitious neighbors, often living right next door (Kennedy 1987, 20–22). Germany and Italy were loose geographical designations for broad areas with linguistic and cultural similarities, but they were politically divided into numerous separate entities. The forces of reformation that swept through the sixteenth century shattered the relatively little homogeneity that Europe previously did enjoy through its more-or-less uniform religion. Perhaps the competition among crowded states and peoples drove Europeans to perfect the means by which they could fight their rivals.

At the same time, at least one element of European geography does not make a great deal of sense, at least where the use of firearms is considered. Much of Europe is very damp, north of the Alps in particular. The reliability of matchlock guns in places like England or the Netherlands must have

A seventeenth-century matchlock musket drill sequence. © Hulton Archive/ Getty Images.

changed from season to season. Why was this weapon system so popular in a climate that it was somewhat ill suited to? Having raised this question, it must also be admitted that parts of the Mughal Empire were far soggier, and they too persevered with early firearms.

When looking at the development of firearms, it is easy to get so carried away with "improvements" that one tends to ride roughshod over earlier guns and neglect what really successful designs some of them were.

The matchlock is a case in point. Like any technology, it had some draw-backs. The match had to be readjusted in the serpentine frequently as it burned down. It was also a bit dicey wielding a burning cord around gun-powder stores. The match glowed and gave off a strong smell that might have been less than helpful while trying to spring an ambush (however, this certainly would not be much of a liability in a large field battle). Fi-nally, it must be admitted that it did not work all that well in wet weather. One important invention that helped the matchlock a little in this regard came from the dreary, rainy Netherlands. It was a cylindrical tin or brass container pierced with ventilation holes. This matchcase was hooked to a belt and could protect a smoldering match. When the arquebus was ready to be fired, the lit end of the match was pulled out and fitted into the ser-pentine. Between shots it was put back into the matchcase. The device gave gunners a place to secure the burning tip and protected the same from mist and drizzle.

With this handy device, even in moderately damp conditions the matchlock may have been even more reliable than later flint-and-steel igni-tion mechanisms. At a historic seventeenth-century reenactment that I wit-nessed, one participant with an early flintlock was having poor luck in the humid summer air. After several unsuccessful attempts at firing his musket, he borrowed a piece of burning match from his arquebus-armed neighbor. Holding the stock in his left hand, he braced the gun against his shoulder and abruptly jammed the burning match end into his weapon's flashpan. There was no further argument from the musket, and it promptly discharged. We might well imagine that such scenes happened historically. Why should any-one have been looking to replace such a reliable ignition system?

THE WHEEL LOCK

Recognizing that the matchlock did work well, like other weapons we have looked at, someone in Europe felt inclined to fix it even though it was not clear that the old design was actually broken. The basic job of the match was to ignite the powder charge of the firearm. Obviously it worked well; how-ever, other methods could be employed to do the same job. As far as we know, the two-sticks-of-wood-rubbed-together-lock or magnifying-glass-positioned-to-the-sun-lock was never attempted. There was yet another an-cient method of creating fire to try. People had known for a long time that it was possible to create sparks by striking steel with a hard piece of flint. This method was apparently employed in a different type of gunpowder weapon even before its use in the firearm. Some evidence exists that flint and steel ignition systems were used to detonate underground land mines in China as early as the mid–fourteenth century (Needham 1986, 199–202).

The next step was to put this application into use as a gunlock. The oldest effort in this direction may have been a fifteenth-century oddity known as a "monk's gun." This weapon had an iron rasplike bar running through the bottom of a boxlike flashpan set parallel to the barrel. The sliding rasp extended out past the flashpan toward the shooter, where it ended in a finger loop. Instead of burning matchcord, the gun's serpentine arm was designed to hold a piece of flint in a jawlike clamp. The arm was fixed in place with the flint resting on top of the rasp. When this was manually pulled backward, it scraped against the flint to create a shower of sparks. These would ignite the priming powder in the pan and discharge the weapon. The idea was apparently not hugely popular, however, and only one example exists today.

Still other people continued to explore the principle of flint and steel ignition systems. The most famous was Leonardo da Vinci, who made an undated illustration of a wheel lock sometime before his death in 1519. Whether he invented the device or just sketched one he saw falls into the realm of speculation. In either event, the lock he drew appears quite technical in its composition and suggests that the basic design was on the scene in the earliest part of the sixteenth century.

The wheel lock was a remarkably clever system based upon a similar principle to that found in the monk's gun. However, the whole device was more mechanically driven and easier for the firer to use. The major difference was that a wheel, the basis of the lock's name, essentially replaced the long rasp. The idea was for the wheel to spin in place while its serrated edge ran against the flint, a device like the modern lighter. The stone was held in a jawlike clamp attached to a hinged arm that was similar in position and function to the serpentine. Just as the serpentine owed its name to an animal, the canine appearance of the structure led it to be called a "doghead."

The shooter prepped the wheel lock by using a separate spanner wrench to wind a lug fixed to the axis of the wheel. This action wrapped a very short (two-link) chain around the wheel as it turned, similar to the way one would wind up a toy yo-yo or a top. After being cranked around, the whole thing was held in place by the nose of a metal sear, which popped into a hole drilled in the side of the wheel. The upper part of the wheel extended up into the flashpan located just above it and below the doghead. When the shooter was ready to fire, the cover was pushed away from the flashpan and the hinged doghead was manually swung down against the top of the wheel. When the trigger was pulled, it allowed the main spring to reassert itself and pull out the chain that, in turn, spun the wheel. This made an approximately three-quarter rotation with its serrated edge spinning against the flint held in the doghead's jaws. The resulting shower of sparks set off the priming powder and main charge.

The wheel lock was a fairly complicated bit of machinery in comparison to the matchlock. In spite of all of the intricate bits and pieces, the basic idea was sound and, when working properly, the weapon's discharge occurred almost simultaneously with the pulling of the trigger. There were, of course, a number of variations to the system outlined above. The wheel could be fixed on the lockplate externally or internally. Most of the latter wheel locks contained a secondary spring system that automatically pulled the flashpan cover back when the doghead was swung down. Others were designed so that the doghead rested directly on a spring-locked cover. Only when the trigger was pulled did the cover snap back, dropping the flint onto the wheel. Some versions had two dogheads or a doghead *and* a match serpentine fixed to the same lock as a backup mechanism in case of misfires.

Early models also required a lot of strength to operate, so some later versions had modified triggers that could be set to go off with different levels of pressure. Setting the pressure too low could lead to accidental shootings if caution was not used. Unsurprisingly, safety-locking mechanisms that helped prevent the weapon from accidentally discharging also were developed. This improvement came too late for a Constance courtesan, who was wounded in the chin and neck on January 6, 1515, in what appears to be one of the earliest written references to both a wheel lock and an accidental shooting (Reid 1984, 67). A further safety measure to appear on wheel locks was a thin metal bar that ran around the trigger. This steel guard offered a degree of protection against accidental discharge or casual damage to the trigger itself. The space on either side and in front of the trigger was open so that the shooter could fit his index finger in when ready to fire. Some trigger guards were long, elaborate designs with notches for the other three fingers of the shooting hand to grasp.

There were certainly some disadvantages to the weapon, even in its improved manifestations. The system was complicated, delicate, and prone to breakage, particularly those with an externally mounted wheel. Internally fixed wheels were more durable, but were harder to clean and tended to clog up with powder residue. A wheel lock also demanded more craftsmanship to make and more money to purchase. The fact that they were not used in huge numbers by common troops may have been a factor of cost as much as reliability. Remember that the matchlock continued to be popular throughout the "lifespan" of the wheel lock. But if the matchlock was still a useful weapon, what advantages did the wheel lock convey?

In at least two applications, it performed impressively. First of all, it may have played at least a partial role in the salvation of mounted troops. Although it was not impossible, wielding a matchlock on horseback was difficult. The problem was not so much the match going out, but rather the constant readjusting of its burning wick. Managing this while maneuvering a horse was bad business. Handling a gun the size of an arquebus while riding could not have been easy either. Enter the wheel lock pistol.

The term "pistol" has been traced to several sources, such as the town of Pistoia in Italy or "pistala," a Czech word for tube. The term "dag" was another common sixteenth-century designation for these short, single-handed firearms. Although the Japanese and Poles are known to have produced some matchlock versions, the development of this particular weapon was largely dependent upon the advent of the wheel lock for the reasons outlined above (Held 1970, 43). In fact, it was said to be the type of gun that wreaked havoc on the hapless courtesan. Pistols could do more than damage the teeth of high-priced female escorts. They now provided mounted troops with the means to shoot back at those arquebusiers and musketeers who were wreaking so much havoc on their ranks. A wheel lock pistol had no troublesome match and could hang from a belt or a saddle, ready for firing at a moment's notice. Typically cavalry troopers carried

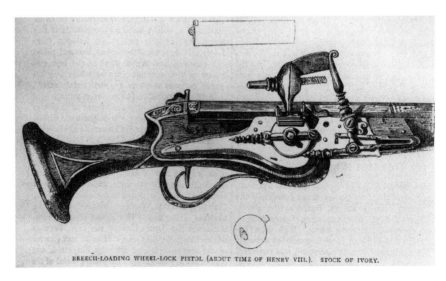

BREECH-LOADING WHEEL-LOCK PISTOL (ABOUT TIME OF HENRY VIII.). STOCK OF IVORY.

A sixteenth-century wheel lock mechanism. © Hulton Archive/Getty Images.

three or four of these guns, two in saddle-fixed holsters while the extra weapons were tucked in a boot, belt, or waist sash. The cavaliers could fire these at the enemy before charging in with a sword for melee combat. Cavalry also developed a complicated maneuver known as a "caracole" in which the front rank of horsemen would fire their pistols and then make a 180-degree turn back to the rear of the formation (Dupuy 1980, 116–117). They would theoretically start reloading their pistols as additional ranks of horsemen moved up in line and began firing. This was supposed to be done in succession to create a continuous volley of shots. While visually impressive, the maneuver had its limitations. The first run-through probably generated a high initial rate of fire, but the reloading process must have been dubious at best. Furthermore, the pistol had a very short range and the formation would have been subject to some infantry fire before drawing close enough to engage. It has been estimated that cavalry may have had to close to within five paces of the enemy in order for their wheel lock pistols to work to full effect (Ellis 1978, 81–84).

Eventually some cavalry units developed other firearm tactics. One technique was to use dismounted cavalry armed with guns as highly mobile infantry. This had been done informally by troops under Count Peter Strozzi using arquebuses as early as 1537 (Childs 2001, 155). The cumbersome arquebus was ill suited to the role, however, and a lighter weapon was needed. There was a vast range of different-sized guns that could be made, and a number of these hybrids fell somewhere between the pistol and the arquebus, both in their design and application. Smaller two-handed guns such as the petronel, musketoon, and carbine could provide horse-mounted troops with a manageable-sized weapon that had a longer effective range than the pistol. By the end of the century, carbine-armed cavalry who had been specially trained to fight as dismounted troops were being regularly raised in France. Other states would follow suit through the course of the next several decades. They came to be known as "dragoons" perhaps in some obscure linkage to dragons. To give them some additional flexibility, dragoons still carried a sword and often a backup pistol or two. Interestingly enough, the development of shorter carbinelike weapons designed for horseback use reduced the hostility toward the firearms amongst some equestrian cultures that had previously rejected guns, such as the hapless Mamluks mentioned in Chapter 2 (Ayalon 1956, 96–97).

Cavalry never regained the predominance it once had enjoyed in the High Middle Ages and was generally employed to scout, skirmish, and run down broken units of infantry. Still, we should not overemphasize these changes too much. Vestiges of knighthood remained, particularly in specialized units known as lancers who relied on shock-attack tactics and continued to be employed successfully as late as the early nineteenth century. Ironically, some late-sixteenth-century dag-armed horsemen were known as "reiters," a term related to the German word "ritter," meaning "knight" (and, of course, more distantly related to the English word "rider"). Flint and steel firing mechanisms like the wheel lock and its successors helped breathe life back into the horse-mounted soldier. Their mission was now somewhat peripheral, but the infantry firearm had not yet completely erased cavalry from the battlefield.

Beyond their military applications, wheel lock pistols had a second role in the realm of self-defense. A musket would certainly protect someone too, but the pistol was much easier to carry and conceal. The wheel lock mechanism could stay at rest for a long time and still be ready to fire at a moment's notice in case of a threat. Imagine trying to walk around town with a smoldering matchlock concealed under one's cape or hooked to a belt. If a person could afford a pistol, they did not have to be physically large or martially trained to enjoy a measure of personal power, for good or ill purposes. The long march to Samuel Colt's "great equalizer" had begun.

One further area of success for the wheel lock needs to be considered. Firearms are not always intended only for the business of killing other humans. At the beginning of this book, we noted that another primary imperative behind the rise of human weaponry was hunting. Thousands of years of agriculture had not eliminated this basic impulse. Hunting in early modern Europe remained a popular sport and a secondary source of food. Matchlocks were employed in this activity, as evidenced by a print from 1566 that appears to show commoners hunting waterfowl with arquebuses (Held 1970, 63). This could not have been easy, as wild animals are not remotely considerate about letting themselves get killed easily, and hunters usually have to wait for long periods of time between shots. Constantly mending a glowing match would have been a major headache in this endeavor. The odor of the burning cord might also have constituted a significant problem. The fact that sporting crossbows lasted as long as they did may also be taken as evidence of these drawbacks. If one could afford it, the wheel lock was better suited to this sort of work. A hunter so equipped was able to hide or stalk for hours without worrying about whether his firearm was ready or not.

These early hunting guns could be loaded with different types of ammunition depending upon the game. A single heavy ball was used to take down large animals like deer or boar. For upland birds, waterfowl, and other smaller animals, a load consisting of little lead pellets known as birdshot was employed. The shot itself was made from lead that was cast, cut from larger pieces, or melted and dripped in small amounts into water. When fired, it would spray out in a wide area, hitting almost anything in this so-called shot pattern. This clearly was a bad development for the aviary world. Shooting even a resting bird with an arrow or a crossbow quarrel (a square-headed arrow used specifically in crossbows) must have been an exceptionally challenging feat, but a blast of birdshot might kill several such animals at once.

Some massive musketlike weapons were designed just for this purpose and became known, appropriately enough, as "fowlers" or "long fowlers." While the weight and length of the guns drastically limited their mobility, one other practical reason for their size involved ballistics. The exceptional length was needed to keep the birdshot clumped together longer as it traveled down the barrel. A short tube would send the pellets scattering in all directions. Contrary to popular mythology, the early-seventeenth-century New England Puritans did not use short, wide-flaring-muzzled blunderbusses to shoot wild turkey. That gun, probably related to the Dutch word "donnerbus," or "thundergun," was more common to the eighteenth century and was intended for short-range use by coach guards and constables in the manner of the nineteenth-century sawed-off shotgun.

While hunting influenced the further development of the firearm, most of the motivation for change continued to be driven by military applications. The entire face of warfare in Europe was in flux between 1500 and 1800. A number of historians have used the term "military revolution" to describe these changes. There is, however, significant argument over the causes, extent, and duration of this movement.[1] Currently, some question whether changes in warfare that occurred over centuries really constitute anything as sudden and decisive as a "revolution" (Childs 2001, 15–17). Recognizing that military technology was not static in the Middle Ages, most historians agree that in general, however, there was a significant transformation of martial tactics and technology in the early modern period.

One important element in these changes was the basic infantry firearm. We have already seen how the matchlock arquebus introduced a number of changes on the European battlefield. New developments in the seventeenth century would further alter both techniques in battle and the design of the firearm. The average barrel remained largely unchanged, but basic infantry stocks became lighter and more graceful. Furthermore, the

---

1. For a review of these debates, see Parker (1988) and Black (1994).

underlying principle behind the wheel lock would now help spawn a similar, but ultimately more successful, ignition mechanism.

THE EARLY FLINTLOCK

There is more than one way to get flint and steel to ignite. When one considers all the various parts that had to work in harmony, it is clear that the wheel lock was a fairly complicated mechanism. There were simpler and cheaper ways to achieve the same result. It is odd that the first popular flint ignition system would be so complex in comparison to later mechanisms. We tend to assume that technology moves from simple forms to more convoluted ones, but this is not always the case, and such was the situation with the wheel lock. The impetus now was to find a system that conveyed the same advantages but was easier to manufacture, to use, and to service.

One of the first steps in this direction was a lock that was apparently named after its mechanical similarity to a pecking rooster. It was known on the continent alternatively as the *snapplas*, *snaphaunce*, or *schnapphans* (linguistically similar to the English "snapping hen"), although "snaplock" probably works well enough as a general description today. There seems to be much confusion over the national origins of the snaplock, as Germany, Italy, the Netherlands, and Sweden have all been credited as the place of origin. The date of origin is a bit murky too, but it appears that Swedish models were around as early as the mid-sixteenth century, although these may have been produced by immigrant German gunsmiths (Lenk 1965, 3–4).

The snaplock appears to have been influenced by a new type of matchlock design that emerged in Germany a decade or two earlier. In this particular version, the serpentine was locked in an upright position and, when fired, literally snapped down into the flashpan. In the snaplock, a similar system was used, only the serpentine was replaced by a doghead holding a piece of flint. This doghead was narrower and more beaklike, hence its association with chickens. It came to be known as a "cock" in some countries, while canine designations persisted in others.

Although there were numerous variations, the snaplock in general was a much simpler mechanism than the wheel lock. In this design, the cock was mounted toward the rear part of the lockplate. It was pulled upward and backward and locked into an upright position prior to firing, an action from which the term "cocking" a gun emerged. Pulling the trigger, of course, caused the cock to fall down and forward, and its flint struck a stationary piece of iron called a battery. Theoretically, the shower of sparks created when the flint scrapped the battery poured directly into the flashpan, igniting the priming powder, which in turn set off the main charge in the barrel.

This system was a little slow to catch on and did not begin to become common until the early seventeenth century. Matchlocks remained popular with the troops, while wheel locks retained a firm grip on the social elite. The wheel lock had a shorter lag between the time the trigger was pulled and when the weapon actually discharged. The snaplock had a slightly longer delay in firing and probably had a higher rate of misfire in general. For individuals of means who could afford it, the wheel lock seemed a more dependable choice. When one considers purchase orders in the thousands, however, even the cheaper snaplock must have seemed prohibitively high. Snaplocks might be employed in the firearms of specialty units or in cavalry pistols, but the matchlock still was the exchequer's weapon of choice for the common foot soldier throughout much of the seventeenth century.

Before the military forces were to abandon their matchlocks en masse, variations and improvements on the flint-and-steel mechanism would continue to emerge. It was a fairly short step from snaplocks to flintlocks. The term "flintlock" today has a relatively specific meaning, but this was not the case in the seventeenth century. The phrase "snaplock" might be employed to describe what we today would call a flintlock or vice versa. If one thinks about it, any spring-triggered firing mechanism using flint is by definition, a "flint lock." Still, writers today argue about what was a "true" flintlock and what was not.

Certainly an early candidate was a late-sixteenth-century Spanish ignition device, later known as the miquelet. This externally mounted ignition system had a special L-shaped combination battery and flashpan cover. This battery was fixed so that the vertical back of the "L" was set closest toward the shooter while the bottom horizontal piece sat on top of the flashpan, acting as a cover. When the cock hit the back of the "L," it fell forward, lifting the horizontal bottom up at an angle from the flashpan and exposing it to the sparks. The miquelet version of the flintlock was an extremely successful design that was still widely used over 200 years later in the Spanish Peninsula campaign against Napoleon.

In the first few decades of the seventeenth century, different versions of this system such as the English lock, dog lock, or patilla lock appeared in Italy, England, and Scotland. At some point, the first "true" flintlock was designed, probably by the brothers Marin and Jean le Bourgeoys, gunsmiths to Louis XIII of France (Lenk 1965, 30–37). This flintlock worked on the same L-shaped battery-flashpan cover as the miquelet, but most of the working parts were fixed internally, behind the lockplate. They were thus better protected and less subject to wear and tear. The lock's sear also had a half-cock tumbler notch. This meant that the cock could be pulled partially back and locked in a "safety" position. If the shooter accidentally pulled the trigger, a properly functioning gun would not fire. The qualification "properly functioning" must be employed because at some point the phrase "going off half-cocked" eventually came into parlance.

Flintlocks admittedly did not fire as quickly as wheel locks, but they did offer certain advantages. For one, they were simpler to use and to service. In particular, there was no winding involved. It was easier to simply cock the weapon and fire, and there was no hassle of carrying a separate spanner that might become lost at an inopportune moment. The wheel lock's complicated mechanism also meant that there were many more parts to potentially break or become gummed up with powder reside. The flintlock's ease of use and convenience were significant enough to lead the wealthy to begin discarding their old weapons in favor of it. The fraction-of-a-second difference in their firing speeds apparently came to be viewed as an insignificant liability.

THE FLINTLOCK MUSKET

What caused the decline of the venerable and cheap matchlock? This is an excellent question for which there may not be an entirely satisfactory answer. Initially, only certain elite units were equipped with expensive flintlocks. For example, the modern British Royal Welsh Fusiliers were originally formed to guard artillery trains. Since their mission kept them in the vicinity of powder stores, they were issued flintlocks in order to reduce the risk of an accident caused by a mislaid smoldering match. Their lightweight firearms were colloquially known as "fusils," a word linked to a Latin term for flint-striking steel. Beyond some exceptions, however, for nearly a century the military ignored the flintlock in favor of the tried, true, and cost-efficient matchlock. The flintlock did eliminate the headache of fooling around with a smoldering match, and thus theoretically at least enjoyed a slightly higher rate of fire. Given the high incident of misfire, however, one wonders how true this really was. Nevertheless, in the 1690s the major armies of Europe quickly began switching to the pricier device. In this age of absolutist rulers, pacifist budget directors were run roughshod over by hawkish sovereigns who had few internal checks upon their ambitions. Fear of *external* checks and the desire to maximize their troops' rate of fire led them to swallow the expense of the upgrade. Not all states were absolutist, the Dutch Republic and Britain being notable constitutional exceptions. But if these two powers wished to remain independent nonab-solutist states, they too had to adapt. Between 1689 and 1703 Austria, Britain, France, the Netherlands, and Sweden all adopted flintlock firearms as the primary weapons of their infantry.

Arquebusiers and musketeers back in the sixteenth century had worn a type of leather strap running across the chest known as a bandolier. Hanging from this belt were twelve or more wooden cylindrical containers holding premeasured charges of powder along with a projectile. Sometimes known as the "twelve apostles," these saltshaker-looking devices sped up the reloading process since infantrymen did not have to waste time trying to measure the load. In 1590, an Englishman, Sir John Smyth, mentioned containers like these that were made out of paper. Their use seems to have predated his description, but it is impossible to say by how many years. King Gustuvus Adolphus of Sweden equipped his troops with these conveniences, and they were widely in use across Europe by the end of the Thirty Years' War. The paper cartridge was opened with one's teeth, and a small portion of the charge was poured in the gun's pan to prime the weapon. The rest of the powder went down the barrel along with the ball. The paper tube could be rammed to pack the charge more tightly, or simply discarded. Troops were able to carry many more of these lightweight paper cartridges in a leather purse (or "cartridge box") than the mere 12–16 shots that hung from a traditional shoulder bandolier.

Although we tend to think of the firearm primarily as a shooting device, some mention of an important accessory is probably warranted. The pike had been a standard element of field armies at least since its successful use by Scottish formations against mounted English knights at the Battle of Bannockburn in 1314. We have already covered the decline of the mounted knight in some detail, but his old nemesis, the pikeman, was also destined to fall by the wayside. As late as 1571, Spanish forces in the Netherlands still employed a ratio of two men equipped with firearms for every five equipped with pikes. By 1601 a similar unit was operating with a ratio of three guns for every one pike (Parker 1995, 154). By the end of the Thirty Years' War, the ratio was more on the order of four to one. In those circumstances where cavalry did opt for a direct charge, it was still useful to have some pikemen around to fend off the attack and save the gunners. Musketeers often carried hand weapons; indeed, we often mistakenly associate them with swordsmanship thanks to Alexandre Dumas's *Three Musketeers*. The infantryman's sword, however, was a relatively poor defense against a lance- or saber-armed cavalryman. What was needed was a handy pike that the gunner himself could use in a hand-to-hand melee.

The bayonet answered that need. Eventually it was recognized that the long, thin firearm itself could serve as a kind of staff for an attachable spear-head. The earliest efforts along this line were fixable blades created in the sixteenth century for use in dispatching wounded game animals. There are also accounts from the Thirty Years' War of desperate musketeers report-edly sticking the handles of daggers into their muzzle barrels to form rudi-mentary spears (Reid 1984, 122). Toward the end of that conflict, French troops were using specially designed blades with a handle made to the same diameter as their firearm's barrels. This simple device, known as a "plug bayonet," became widespread in the next few decades. The word "bayonet" itself is sometimes linked to the town of Bayonne, where the device report-edly was invented.

One problem with the plug bayonet was that it prohibited shooting while fixed in the end of the gunbarrel. Putting the blade in and pulling it out whenever cavalry threatened to charge could lead to chaos and (even worse) a reduced rate of fire. Some tried to reduce the hassle of this problem with a hinged blade designed to swing back along the barrel when not in use. Hinged bayonets like this could still be found on a number of rifles in the twentieth century, including some versions of the Russian AK47 assault rifle. A more common solution was the simple socket bayonet. The butt of this blade did not end in a handle but rather, as might be guessed by its name, in a socket. This open-ended steel tube was designed to slip over the muzzle with-out blocking or plugging it up. The socket could lock onto a projection on the barrel or to the front gun sight itself. The new bayonet had a sharp S-curve so that the blade was offset from the muzzle by about an inch, and thus did not interfere with shooting. If a cavalry unit were nearby, a regiment could fix bayonets and continue to pepper away at them even if they appeared ready to charge. Now the fifth of the infantry that had been devoted to anticavalry guard duty could be replaced by more usefully armed troops.

It should also be noted that the typical firearm had shrunk in size over the previous century. The original massive musket disappeared, and its name was transferred onto those lighter types of guns that earlier had been known as calivers, a name that, like "arquebus," faded into history. Stocks grew less heavy, less cumbersome, and less (pardon the pun) "stocky." These guns were small enough to remain fairly agile even with a bayonet fixed on the end. The change in size was probably related to improvements in firearm design. Gunsmiths won the centuries-old competition with armorers, and most mounted warriors abandoned plate mail by the mid-seventeenth century. A massive firearm like the old version of the musket was not needed to pierce a linen waistcoat. Like the lance, helmets and breastplates remained popular with certain cavalry units for another two centuries, but they were intended to offer protection against sabers, not against shooting.

Since a variety of firearm sizes was no longer needed, armies were able to be more demanding about the standardization of weapons. This was particularly true with regard to caliber sizes. The obvious advantages in terms of logistics and shared ammunition had been recognized for a long time. At least as early as 1599, Maurice of Nassau had equipped units with the same caliber guns as part of a series of military reforms. Like other innovations noted above, this pattern accelerated during the course of the century. Variation existed between individual countries, of course, but serious efforts were made to use a standardized caliber size for basic infantry weapons. Usually this was somewhere around ¾ of an inch (.75 caliber) in diameter.

By the end of the seventeenth century, England was having mixed success in moving toward this goal. They had managed to adopt a standard caliber size, but their nicely uniform ammunition was being thrown from a hodgepodge of different weapons of uneven quality. As a result of the uneven performance of British firearms during the War of Spanish Succession (1701–1714), the duke of Marlborough led the effort to develop a superior-quality, standard weapon for his infantry. The end result of his efforts was the Long Land Service musket, also known as the King's musket, Tower musket, or the "Brown Bess." The third appellation came from a Tower of London stamp inscribed on the lockplate, while its more colorful informal name came from a chemical treatment that was applied to the steel parts. This was intended to ward off rust and left the metal a brownish color. God alone knows whom the "Bess" came from, although much speculation exists.

This weapon was one of the most successful designs in the history of firearms and can be used as the prototypical example of an eighteenth-century infantry musket. Its service life actually extended well into the following century, with versions still being cranked out in Birmingham as late as 1842 (Greener 1967, 624). The design was fairly simple: it had a 46-inch-long, .75-caliber barrel fixed to a walnut stock, and in later models some of the "furniture" was made of brass. This last term refers to decorative but functional niceties like the trigger-guard or the "butt-plate." The latter was intended to reduce wear and tear on the rear of the stock. There was a bayonet lug at the muzzle that did function as a sort of front sight, but there was no corresponding rear sight, a testament to the gun's lack of long-range accuracy. Later proponents of precision rifles were quick to belabor this weakness, but like most muskets of its time, the thing was not really built with an eye to individual marksmanship. The idea was for dozens of these weapons to be fired at once at mass formations of troops, and for this role it was well suited. It also met the ongoing imperative for a high rate of fire, since it was easy to shoot and reload. While some experimentation back and forth with wood occurred, most models wound up using an iron ramrod, an accoutrement that was largely standard on the majority of continental European flintlock muskets by 1750. This metal accessory was less likely to break under duress, although generally muskets did not have much of a problem with this for reasons that will be reviewed in the next chapter.

Besides ramrods, furniture, and other minor alterations, the Brown Bess underwent a gradual shrinking during its long tenure. Around 1750 the Short Land Service variation, with a 42-inch barrel, was developed, followed by the nearly identical New Pattern Short Land Service model in 1768 (Brown 1980, 228–231). Tests showed little performance difference between the 46- and 42-inch barrels, and the latter were slightly cheaper and easier for dragoons or ship-based marines to handle. The financial and material demands of the Napoleonic wars led to a third major design with a 39-inch barrel, based on the East India Company's preference for an even shorter weapon. Known simply as the India, or Third Model, thousands were produced by independent gunsmith firms.

The British Brown Bess was the exemplary flintlock musket of the eighteenth century, but other nations produced similar weapons. French muskets were named after the year they were adopted, such as the Model 1717, Model 1728, or the Model 1746. Like the Brown Bess, these weapons picked up informal nicknames such as the "Charleville" after the royal arsenal in that town. Prussian troops used a musket known informally as the "Potsdam," and many Russian soldiers sported a "Tula" for similar reasons.

From circa 1680 to 1830, warfare was dominated by these musket-armed infantry. Sieges and heavy artillery remained a key element of combat since more attacks on cities and fortifications occurred during this period than did large-scale pitched battles. But even in a siege, firearms played an important role. Harassment along trench lines and the eventual rush through breached walls required the musket and bayonet. Furthermore, many military encounters of the seventeenth and eighteenth centuries were skirmishes and small-scale affairs for which the light flintlock musket and bayonet were well suited. The handheld firearm was now beginning to enter into the peak of its game.

In combat, soldiers were deployed in very long and thin formations. The term "line infantry" thus came into use for obvious reasons. Now, instead of firing in rotating ranks of six or ten men, troops were deployed in lines of three or even two. A common formation was to have the front rank kneel, the second rank crouch, and the third rank stand. This technique was described as far back as the Thirty Years' War, but was not universally adopted until much later. No rotation was conducted; instead, the idea was to save time by standing still and reloading as quickly as possible.

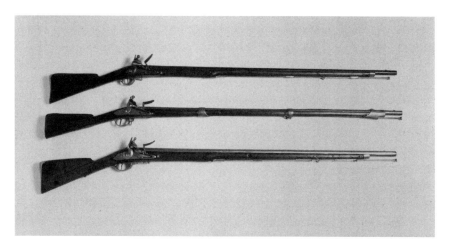

Three types of flintlock muskets, including a Charleville (center) and a Brown Bess (bottom). © Hulton Archive/Getty Images.

If a commander wished to keep up a continuous "rattle of musketry," he could order his troops to fire in small groups called "files." For a bigger bang, an officer might have his troops fire by rank, with each man in a rank firing simultaneously while the other two reloaded. Finally, for the greatest single psychological blast, all three ranks of the entire unit could fire a single simultaneous volley.

The firearm was hitting its stride by the late eighteenth century. It had moved from being a battlefield accessory to occupying a central place in the calculations of military leaders. Much of this had to do with changes in its performance and lethality. The dangerous matchlock arquebus gave way to the deadly flintlock musket, which dominated the relatively open battlefields of Europe. Like the matchlock, the flintlock would also travel around the world and see action in a variety of places and conditions. Despite its apparent success, however, political and geographic conditions would come to conspire against it, leading to yet another chapter in the life cycle of the modern firearm.

4

Rifling: A New Twist on an Old Idea

◆

THE DEVELOPMENT OF RIFLING

Like the matchlock arquebus, the flintlock musket was another exceptionally successful design. It too was invented in Europe, but versions of it appeared in many places around the world. The basic system was often imitated (and easily duplicated) in local shops from India to North America, often with an eye toward local artistic designs and traditions. Recent research shows that the flintlock musket made a strong impact in areas of the world traditionally presumed to have lacked all but the most rudimentary weaponry. By the early eighteenth century, large armies of West African states were armed entirely with flintlock muskets, much like their European counterparts (Thornton 1999, 61–64). There was obviously a wide variation in the quality in these weapons, and few regional craftsmen could meet the exacting standards of a late-model Brown Bess. Still, the basic design enjoyed a widespread reign as the world's predominant gun design for about a century and a half. Like other stages in the life of the firearm that we have reviewed so far, the flintlock musket adolescence would likewise come to a close. The gateway to adulthood is imprecise, however, and the weapon did not vanish overnight. Two of the firearm's vital components, the barrel and the lock, would drastically change by the mid-nineteenth century. This chapter will consider the first of these alterations.

Returning to that somewhat obvious (and admittedly pathetic) technique of sports analogies, the manner in which a North American football player throws a ball warrants some attention. The quarterback grasps the projectile with his or her four fingers on one side and thumb on the other. If the throw is made properly, the thumb will release first, followed by the other four fingers, which roll more gradually off the ball as the arm swings forward. This rolling motion will cause the ball to spin in midair, increasing the range and accuracy of the throw. The technique takes a little practice, and for many of us playing in the backyard, the common result is a wobbly, end-over-end lob more likely to hit the neighbor's dog than the arms of an open receiver. Those who are really able to master this skill, however, can make throws of mind-boggling precision and go on to earn even more mind-boggling salaries.

The ball performs better due to its spinning action. Physical imperfections, such as the ball's strings, become more equalized in their blurring trip around the central axis. The spin essentially adds an artificial element of balance and stability, much in the manner of a gyroscope. Although the physics behind the effect were not understood, the basic technique has apparently been known since ancient times and was frequently employed to make arrows travel in a more accurate manner. The fletcher (arrow maker) would simply attach the feathers to an arrow's shaft at a slight angle, which would then cause the device to pick up a spin as it left the bow. In the more fundamentalist sixteenth century, the presence or absence of demons on a spinning missile was suspected to have played a key role (Held 1970, 36, 138–139). A more legitimate explanation for the phenomenon was to remain absent, however, until an Englishman, Benjamin Robins, accurately explained the principle in the 1740s.

Somehow the technique was passed on to firearms during the initial burst of experimentation that occurred in the early modern era in Europe. There are two possible causes for this development, one accidental and one intentional. There are different versions of the story but according to some accounts, rifling was first developed to address the common problem of powder residue fouling the barrel after several shots. The attempted solution was a trench cut into the interior sides of the barrel into which unburned residue could settle. A number of weapons with *straight* grooves in the barrel exist, which clearly lends credence to this theory. For some unknown reason, a gunsmith once cut the grooves in a curved, spiraling manner, and it was discovered that balls shot from such barrels traveled in a more accurate fashion. Given the fact, however, that arrows were made to fly in such a manner, it seems at least equally likely that someone intentionally cut spiral grooves into a barrel in hopes of achieving a spinning effect. After all, the firearm borrowed many of its basic parts from the crossbow, so why would gunsmiths not also seek to borrow the flight patterns of its projectiles?

Whether by accident or by design, rifling probably began to appear in the last quarter of the fifteenth century. The oldest known example is a bronze weapon from the collection of the German Emperor Maximilian I, which carries the coat of arms he used from 1493 to 1508. Rifles were produced all over Europe, but they were particularly popular in Germany. There may be a culture reason for this geographical exclusivity, as will be reviewed below.

A rifled gun could employ any type of lock. Maximillan's weapon, noted above, is believed to have originally carried a matchlock, but later rifles tended to use wheel locks and flintlocks. In addition to the variety of firing mechanisms used, there was a real diversity demonstrated in the rifling itself. The number of actual grooves cut in the barrel might be anything from three to twelve. The number of times (or rate of twist) the grooves turned around the inside of the barrel could also differ from gun to gun and even vary within a single barrel itself! This could not have been easy; forging a plain smoothbore barrel was long and arduous work, let alone making one with spiral grooves cut into it.

For this and other reasons, rifles were expensive weapons that remained in the hands of a wealthy elite in the sixteenth and seventeenth centuries. The earliest full military unit known to be equipped with these accurate weapons was the Danish royal guard, who were given them around 1600. Louis XIII of France armed his bodyguard in a likewise fashion and added two (yes, only *two*) riflemen to each of his light cavalry regiments. The German states of Hess and Bavaria also had small, elite units that were similarly armed. Rifles were slow to catch on amongst most military thinkers and would see relatively little combat before the late eighteenth century. Obviously, a large part of the reason for this was cost. As noted above, sixteenth- and seventeenth-century courts were initially reluctant to even purchase flintlocks due to their price, and were not eager to buy thousands of expensive rifle barrels. Nevertheless, by the dawn of the eighteenth century, most European states had finally decided that military security outweighed financial security and adopted the flintlock musket. Why did they not view the rifle with a similar level of urgency?

The answer again lies in that ancient imperative: rate of fire. In order to get the projectile to spin, it had to be slightly larger in diameter than the basic bore so that when loaded the extra lead would fill, or "grip," the grooves. This was a miserable process as an oversized bullet had to be roughly jammed down the barrel by the ramrod, often with the assistance of a wooden mallet. By the time the ball reached the powder charge at the end of the barrel, it was less a ball than a misshapen, mashed-up hunk of lead. Obviously such weapons had a very low rate of fire. One partial solution (devised in Renaissance Germany, *not* colonial North America) was to use a slightly smaller ball wrapped in a greased patch. Theoretically, such a slippery load would slide down the barrel more easily while the cloth patch gripped into the grooves. As almost anyone who has rammed such a charge will tell you, however, the system sounds a lot easier in the pages of a cozy library book than it really is in actual practice. This is particularly true after a shot or two begins leaving residue in the barrel. The process undoubtedly became even more challenging if undertaken while somebody else's bullets were whizzing by in *your* direction.

For speed of loading, the flintlock musket remained the champion. A soldier was typically issued ammunition that was smaller in diameter than his weapon's bore size in order to make the gun easier to load. Thus the British Brown Bess had a .75-caliber barrel but fired a .69-caliber ball. The .06-inch difference was enough to allow dramatically faster loading than was possible with a rifle. But it also admittedly allowed for a fair degree of gas to slip past the projectile, reducing its force and muzzle velocity. A smoothbore musket clearly had a significant degree of windage. W. W. Greener, member of a famous English gunsmith family, noted this phenomenon in 1841. When describing muskets, he claimed that "the immense escape of explosive matter past the ball prevented the possibility of any velocity worthy of the name being given to the ball, and the range is the most contemptible of any kind or description of gun I know: 120 yards is the average distance at which the balls strike the ground when fired horizontally at five feet above the level" (Greener 1967, 624). Since muskets lacked accuracy at long ranges anyway, military planners traditionally considered the loss of propellant force to be worth the trade-off for rapid reloading. With a musket it really was possible to get off several shots in a minute, while several minutes a shot was more likely for the rifle. That expert on rifling, Benjamin Robins, argued that the improved accuracy would be worth the cost and effort of adopting rifles, but he remained distinctly in the minority.

Although much of the firearm's story is linked to warfare, the rifle found a more important role in the field of hunting. As noted previously, rate of fire is not a particularly important factor while pursuing most game. The crossbow was thus better suited to this role than was the longbow. The rifle's slow rate of fire was really no handicap at all when stalking big animals. One was likely to only get one shot at a swift stag anyway and it was better to be more certain to hit the animal on the first shot, particular in the case of dangerous quarry. A gun's bore might play a key role in whether the boar got floored, or the hunter got gored. This helps explain the relative popularity of the rifle in the southern German states that retained more semideveloped and wilderness areas than did either northern France or southern England. The high cost of craftsmanship that went into the rifle was also less of an obstacle for the wealthy. Outside of poaching, one had to be a landowner to hunt large game in the first place. Ownership of property usually equated to a certain degree of wealth. Furthermore, hunters were not buying hundreds of rifles to equip large military units.

Smoothbore guns remained popular amongst more plebeian hunters and poachers. The rifle was clearly superior when firing a single projectile, but tended to send birdshot all over the place. A smoothbore only worked at short range, but was able to fire either single-shot ammunition or birdshot and thus provided a bit more flexibility for the gun owner. A large number of guns, like the long fowler noted previously, were designed specifically with an eye toward bird hunting. Initially these weapons were heavy devices designed to fire a single cataclysmic blast into a flock of birds feeding, resting, or even sleeping on the water. By the early eighteenth century, more finely crafted weapons with shorter and stronger (usually Damascus-twist) barrels had emerged, initially in Spain. These guns were light enough to be aimed rapidly and allowed shooters to hit birds even in midflight. This "shooting-flying" became something of a craze in the early eighteenth century.

The popularity of this type of hunting may also have been linked to a relative decline in larger game in Europe. For centuries redundant and contradictory hunting laws were passed to preserve game, and while these regulations could be quite Byzantine in nature, so were the penalties of breaking them. Fines, destitution, imprisonment, and even execution might be the result of poaching. An infamous case of uneven justice from the early nineteenth century involved Britain's Lord Palmerstone, who once defended a disgruntled veteran that had shot and wounded him in a botched assassination attempt. Four years later, this part-time humanitarian did not offer any aid to a second man who was convicted and hanged for poaching on one of his estates. It was one thing to shoot Lord Palmerstone, but it was apparently a much more serious matter to shoot his game.

If hunting remained largely the preserve of the landed classes in Europe, little restrictions existed in frontier areas such as could be found in the New World or Russian Asia. Hunting in these areas was customary for both the indigenous populations and the European settlers. Often the newcomers brought weapons from their homelands. Early Dutch settlers of the seventeenth century lugged long, smoothbore fowling pieces from the wet, swampy Netherlands to their settlements on the Hudson and the Delaware to take advantage of the migratory flyways along these rivers.

In the early decades of the following century, Germans began to move into Pennsylvania, bringing their rifles with them. The German word for hunter is "Jaeger," and these relatively short, heavy-bore weapons were known as Jaegers, or Jaeger rifles. The so-called Pennsylvania Dutch began to redesign such weapons to meet the challenges of life in frontier America. By the mid-eighteenth century, the obsession with large projectiles began to disappear and the bore sizes of the rifles dropped from the standard European ¾ inch down to around .60 caliber. This reduction is sometimes claimed to be the result of limited supplies of lead available in North America, but contrary to such a line of thinking is the fact that larger caliber smoothbore muskets still remained common in most of the colonies. On the other hand, the smaller ball size meant that more shots could be carried around per hunter. Perhaps to compensate for the smaller bore size, these guns began carrying longer barrels that grew from around 3 feet in length to 4 feet or more. The extended barrels caused the powder to burn more fully and increased the muzzle velocity at which the bullet was propelled, thus allowing a smaller bullet to become more lethal. More on this phenomenon will be discussed in Chapter 6.

The new German-American gun was initially known as the Pennsylvanian Rifle for obvious reasons. Its presence was originally limited to the middle colonies, but in the late eighteenth and early nineteenth century the firearm traveled westward over the Alleghenies, where it became even more refined. The barrel grew slightly longer still and the bore became much smaller, dropping down to a mere .40 caliber or less! This later version is generally known by the more famous moniker of "Kentucky Rifle." Even if common legends about trappers shooting the eyeballs out of squirrels at 3 miles are discarded, it must still be acknowledged that these rifles were exceptionally accurate, even by modern standards. One fairly credible report suggests that a company of experienced riflemen could consistently hit a 1-foot square target at 150 yards (Brown 1980, 337).

THE RISE OF THE MILITARY RIFLE

While the rifle was popular with some American frontiersmen, for military purposes it remained largely on the periphery. The flintlock musket continued to be predominant, but some changes in thinking amongst the English occurred on this subject as a result of colonial warfare in North America. We can imagine that local militia would turn out for muster armed with a variety of firearms, including some rifles. These oddities were observed by British troops in the Seven Years' (French and Indian) War, and after the fighting ended, some veteran officers called for the development of a military rifle. The guns made an even greater impression during the American War of Independence. Early in that conflict, Congress authorized the raising of several companies of riflemen from the mid-Atlantic states. At the siege of Boston and the later encirclement at Saratoga, they made an enormous impact upon the British officer corps. The unfortunate General Simon Fraser and numerous other junior commanders were picked off by rifles, and the weapon subsequently came to assume a much more forbidding presence in the minds of the officer class than it had on the actual battlefield. These hunting guns had not been designed to utilize a bayonet and this, combined with their slow rate of fire, meant that riflemen were quite vulnerable to aggressive British line infantry. When they were forced into a hand-to-hand engagement without the support of musket-armed troops, they could suffer high losses. Irregular rifle troops were useful for harassing enemy infantry and for skirmish operations in rough terrain, but were unable to seriously contest well-defended towns and other strategic positions.

As suggested above, however, the role the rifle played on the morale of British commanders was significant. To some extent, the gun inadvertently wound up acting a bit like some modern terror weapons, more capable of spreading fear than inflicting high numbers of actual casualties. His Majesty's Army responded in kind to this threat by hiring perhaps as many as 4,000 Hessian and other German light-infantry troops armed with Jaeger rifles (Brown 1980, 339). St. George Tucker, an American officer, recorded his surprise at the amazingly accurate rifle fire of these Jaeger troops at the siege of Yorktown. While one generally imagines rebel riflemen picking off British troops in the American Revolutionary War, in this particular case Ansbach Jaegers shot half a dozen of Tucker's skirmishers, in the dusk, at a distance the Americans thought was well out of range.

The British also experimented with their own rifle-equipped soldiers. Prior to the onset of hostilities, one officer, Patrick Ferguson, had anticipated the usefulness of the rifle on the battlefield and began designing his own particular version of one. His innovative weapon could be loaded through the rear using a clever system originally devised in the late sixteenth century by Freiherr von Springenstein of Munich. A similar design was later patented in 1721 by a French exile living in London named Isaac de le Chaumette (Hogg 1996, 31). The basic idea was to have a section (or "block") of the breech drop down, thus exposing the rear of the barrel and allowing a cartridge to be inserted. This would alleviate the burden of ramming a tight-fitting bullet and greatly ease the loading process.

Breech loading was not a new idea in the eighteenth century, and gunmakers had been experimenting with it since the late Middle Ages. The

Fanciful image of a rifle-armed American marksman shooting British redcoats. Some-how the woodsman has reloaded his rifle quickly enough to pick off a second soldier, all while apparently defying gravity. © North Wind Picture Archives.

primary problem was manufacturing tight-fitting components that prevented gas leakage from the rear of the weapon. This escaping gas could reduce muzzle velocity and had the potential to cause the gun to burst apart. In general, solutions to the problem on a mass-production scale would have to await the technical advances of the Industrial Revolution. Le Chaumette's particular effort to address it involved a breech-opening system that could be tightly screwed in, but this mechanism tended to clog up after use. The Ferguson system had slots cut through the screw threads for powder residue to fall into, perhaps based on the old idea that grooves could be used to prevent fouling. The breechblock itself was unscrewed downwards by means of the trigger guard that doubled as a handle. Because the screw mechanism used a lot of threading set at an exceptionally high angle, the rear of the barrel could be fully opened with relatively little rotation.

King George III of England personally witnessed a demonstration by Ferguson of his new rifle at Windsor Castle in September 1776. The inventor reportedly discharged seven shots in a single minute, an impressive enough feat to win the Crown's approval to equip 100 troops with the weapon. Along with their new rifles, these men were issued green uniforms similar to those worn by the Jaeger companies. In many ways, Ferguson's experimental "Corps of Riflemen" had more in common with the weapons and tactics of the twentieth century than it did with those of the eighteenth. Almost exactly one year after the Windsor demonstration, Ferguson's unit saw action at the Battle of the Brandywine. In this decisive British victory, his troops performed well and won praise from General William Howe (Pope, *Guns* 1969, 135–136). Because they covered a large assault, however, they suffered high casualties and Ferguson himself was badly wounded. For unclear reasons the unit was not reinforced, and their leader was reassigned to the southern theatre. Although he continued to use rifles in his command, his troops were surrounded and defeated by a superior force of frontier riflemen at King's Mountain in 1780. Ironically, Ferguson himself was shot eight times during the battle and killed, probably by rifles.

After the war the new U.S. government went on to adopt the trusty French Charleville smoothbore musket as its official infantry firearm and began producing copies at a government armory in Springfield, Massachusetts. In the old tradition of naming a weapon after the arsenal in which it was produced, these guns gradually became known as "Springfields." The fact that regular American armed forces largely ignored rifles after the Revolutionary War also implies that they were relatively ineffective on the battlefield, despite what later mythology would suggest.

Although the Ferguson gun disappeared, the British kept up an interest in the military use of rifles. During the Napoleonic wars, a more serious effort to develop a rifle corps was undertaken. Ezekial Baker developed the weapon of choice in 1800 based upon the Jaeger design. Unlike the Ferguson, Baker's rifle was a more traditional muzzle-loader originally intended to fire the Brown Bess's standard .69-caliber ammunition. It was later lightened up to a .612-caliber ball, which used a less heavy barrel but was considered adequate nonetheless. Like the Jaeger, the Baker was a relatively short weapon with a barrel measuring a mere 2.5 feet in length. Unlike most rifles, this weapon was designed to hold a bayonet of sorts. To compensate for the firearm's short stature, it was given a special long "sword" bayonet that could double as a single hand weapon if needed. The use of this strange device led to the 95th Regiment of Foot's unique and amusing command: "Fix Swords!" The Baker rifle used a greased patched ball, but contrary to the fictional work *Sharpe's Rifles*, the maximum rate of fire that Baker-equipped troops could achieve was about one shot a minute (Shepard 1972, 89).

The British raised a number of rifle regiments during the Napoleonic wars and used them to good effect in the Iberian Peninsula and in North America. Like other rifle units, the troops wore green for camouflage purposes and were trained to operate as light infantry. The British were not alone in experimenting with rifle units during these wars. Perhaps because they also had an extensive wilderness frontier, the Russians in particular had come to recognize some of the advantages of rifles. During the conflicts with Napoleon they began to equip regular units with a Jaeger-like rifle, only on a much more massive scale than Britain or any other state. Between the years 1803–1812, approximately *20,000* so-called Tula rifle muskets were issued to Russian troops (Held 1970, 152). Clearly, weather was not the only thing the French had to struggle with on the steppes.

Rifling could be employed in other weaponry besides long arms. In the late eighteenth century, muzzle-loaded flintlock pistols were approaching a zenith of craftsmanship. Like the musket, speed of reloading was considered essential in a military pistol. The situation was somewhat different when it came to "civilian" usage. Pistols with rifled barrels could be employed in target shooting, for self-defense, and in the nefarious fashion of dueling. In these situations, fast reloading was less important or practical than an accurate initial shot.

English gunsmiths designed a number of pistols that could be loaded at the breech, thus avoiding the difficulties of rifled muzzle loading. Earlier in the century some models were constructed so that the entire barrel could be unscrewed and rear-loaded. These weapons worked well enough for a time, but after the parts became worn they tended to explode at the breech. Toward the end of the century, a few models based on Ferguson's more successful design were built, but for the most part pistol makers stuck to the tried-and-true method of muzzle loading.

Demand for quality products was quite high due in part to the rise of pistol duels amongst "gentlemen"—a particularly contradictory term in this period. Dueling was an ancient tradition, but it now began to employ the more modern, fashionable firearm instead of the seemingly medieval sword. Despite the pretensions of gentility that surrounded pistol duels, the gun was really more egalitarian than was the rapier. Fencing was a serious art form that required a fair amount of practice and technique. On the other hand, almost any lazy idiot could lift a pistol and pull its trigger.

Dueling pistols came in matched pairs, of course, although more common weapons made for defense or ceremonial purposes often came in a similar twin set as well. The true dueling pistol was made for very grave purposes, and performance expectations were high. The guns had to have a "light" trigger requiring relatively little pressure to pull, since yanking a heavyset trigger could throw off one's aim. It was also imperative that their ignition systems worked consistently. The flintlocks installed on these pistols were so precisely made and produced sparks so dependably that they were able to make do with very tiny and delicate flash pans filled with just a touch of powder. These guns likewise were exceptionally well balanced and fit the hand smoothly. By the turn of the century, a number of these pistols featured "saw" handles with a slight wooden lip on the pistol grip designed to rest on top of the hand and thereby promote stability.

The standard distance of a pistol duel was twenty paces. While handheld pistols are inevitably less accurate than long guns, rifled pistols at this short of a range were murderous. In England and the United States, a (comparatively) sensible tradition of using smoothbore dueling pistols emerged. This did not prevent expert gunsmith Joseph Manton from producing a line of weapons using very light "scratch rifling" that was difficult to detect from the muzzle but that conferred an exceptional degree of accuracy (Neal and Back 1966, 10, 20, 22). On the Continent, smoothbores were considered cowardly and rifled pistols were the norm. It is perhaps no surprise that sword dueling, with a variety of rules for light wounds to satisfy "honor," retained more popularity there. Still, pistol dueling managed to rack up an enormous cost in human lives in the early nineteenth century. Because dueling was actually illegal in most countries and undertaken in a circumspect manner, it is difficult to get a general estimate of these "honor" casualties. Nevertheless, the roster of famous individuals involved in pistol duels in the late eighteen and early nineteenth century is quite extensive and includes Aaron Burr, George Canning, Earl Cardigan, Giacomo Casanova, Lord Castlereagh, Stephen Decatur, Alexander Hamilton,

Andrew Jackson, Ferdinand Lassalle, Mikhail Lermontov, William Pitt the Younger, Alexander Pushkin, and the duke of Wellington. Pushkin ironically included a fictional account of a pistol duel in his masterpiece *Eugene Onigen* that anticipated some elements of his own death. The lack of seriousness with which conventions against dueling were regarded is underscored by the fact that both Pitt and Wellington fought their duels while serving as prime minister of Britain! Gradually sanity reasserted itself and the practice entered a general decline after the 1830s.

Although much human blood was spilt by gunfire in the early nineteenth century as a result of duels, on a happier note there was a decline in military combat after 1815. Nationalist uprisings and economic agitation did create some violence, but the major multistate warfare that had characterized so much of Europe's previous history was largely absent in the wake of the Congress of Vienna. Given this lack of demand for weaponry, one might expect a certain level of stagnation in weapons development. However, this was not the case, as the technological revolution that was creating an industrial world would also infect the development of weapons. Firearms would come to be affected in many ways, but at least part of the initial surge in technology was directed at rifling.

THE CHALLENGE OF THE SLOW-LOADING RIFLE

The successful use of rifles in the Napoleonic conflicts lead early-nineteenth-century military experts to reconsider whether the weapons might not see even more active service, or be issued generally to all troops. In the face of massive war debts, the cost of reequipping tens of thousands of soldiers with new weapons was initially a major concern, but throughout the course of the century fear of other countries again came to outweigh fear of financial ruin, and military expenditure and experimentation began to grow again. A second major sticking point for the general adoption of the rifle was that ancient imperative: rate of fire. Rifles remained pathetically slow to load in comparison to the trusty musket, and troops armed with the former were seriously outclassed by troops armed with the latter at short ranges. Until this problem could be overcome, the rifle would remain an auxiliary weapon while the smoothbore predominated.

This whole situation had a key effect upon the world's geopolitical situation. As long as the smoothbore was the key firearm of the West, there was a clear technological roadblock to European world hegemony. The Americas had admittedly by now been largely overrun, but this had more to do with the devastating effects of transplanted diseases than with the superiority of Dutch, English, French, Portuguese, or Spanish soldiers. Native American populations were decimated by epidemics unconsciously brought westward across the Atlantic (Crosby 1972, 35–42). The most lethal weapons the Europeans used in this conquest were viruses and bacteria, not shot and powder.

The situation in sub-Saharan Africa ironically was almost the exact reverse of that which existed in the Americas. African states did obtain smoothbore firearms from Arab and European traders, but their defense was better assisted by disease. European susceptibility to the seemingly omnipresent malaria helped ensured that their presence on the continent remained peripheral. In the early decades of the nineteenth century, the yearly death rate among European military personnel stationed along the coast of West Africa routinely climbed well above 50 percent (Headrick 1981, 62–64). It was for good reason that Africa was informally referred to at this time as "the white man's grave."

In Asia the environment was less hostile, but total domination still eluded Europe. Western battlefield tactics, diplomacy, and control of international trade had created mercantile empires, particularly in the Indian subcontinent. But these were informal coastal entities with limited influence over interior areas and peoples. The European short-range, smoothbore musket did not really confer much of an advantage over indigenous states defended by the only slightly more archaic weapons. Five or ten thousand average-trained troops armed with matchlocks were more than a match for several hundred of the most highly trained grenadiers armed with the finest Brown Besses. Mercantile empires could also not afford to send thousands of line infantry to the tropics and were forced to rely upon local levies and indigenous allies for support. The general technological firearm parity that existed in the Old World prior to about 1850 imposed serious limitations to the kind of empire building Europe could do in Asia.

This equilibrium would change with dreadful consequences once Europe's new industrial and technological energies were directed at weaponry. One of the first obstacles to be attacked was rifling's slow rate of fire. A notorious failure was the poorly named "Lovell's Improved Brunswick Rifle." While the Baker had served adequately, like all rifles it was difficult to load and fouled quickly after several shots. Mr. George Lovell, inspector of small arms to the Board of Ordnance, began shopping around for a replacement weapon in the 1830s. He was impressed with a Jaeger-like rifle being used by Hanoverian and Brunswick troops. Although Lovell modified the weapon somewhat, the basic version had originally been designed by an officer in the Brunswick army, hence the gun's official moniker.

The Brunswick tried to solve the problem of slow loading by changing both the grooves and the ammunition employed. It seems to have been in imitation of an early Spanish design from 1725 (Hicks 1941, 62). The weapon reduced the number of rifle grooves in the barrel, from the seven found in the Baker down to only two. These two grooves were also cut much deeper than those found in other guns. Furthermore, the Brunswick was designed to fire a special type of ammunition molded with an extra strip of lead running around the waist of the round ball. The idea was for this raised "belt" to fit neatly and smoothly into the grooves cut in the barrel. Theoretically the patched ball would not be scraping across rifling on the way down the barrel, but would rather rotate in the grooves and thus load more easily. The design was also intended to avoid "lead fouling," the buildup of metal particles in the barrel that were scraped off the bullets as they entered and left the gun. Lastly, it was expected that wear and tear on the rifling in the barrel itself would be even further reduced because the ball would travel in the grooves and not across them.

Apparently the extensive tests that Lovell and the Board of Ordnance conducted were promising and the rifle performed very well. Like so many other firearms, once produced in significant numbers and subjected to field conditions the picture became less bright. Because lead is soft, the belts tended to distort and lose their shape from general handling prior to firing. They often did not fit properly and wound up being scraped across the grooves as they went in and out of the barrels. The extra depths of the rifling and excess soft metal on the balls actually increased the amount of friction and lead particle buildup. The grooves also fouled very quickly, perhaps because being deeper they allowed more powder residue to settle and escape being discharged. When fired, the oddly shaped balls tended to wobble in the air, making the gun very inaccurate at longer ranges. The army tried to solve some of these problems by manufacturing special compressed lead bullets, but these may have only increased wear and tear on the rifling itself. Lastly, the gun was a tad heavier than other contemporary rifles. On the plus side, it was one of the first standard military weapons to employ the superior percussion ignition system, but more on this will follow in the next chapter. On the whole, the Brunswick episode further confirms that the development of the firearm did not always follow steady, logical progress, but like most technologies it was punctuated by mistakes, setbacks, and flawed assumptions.

By the time the notorious Brunswick bullet was conceived, more fruitful experiments with superior rifle ammunition were already underway. These would ultimately solve the problem of slow muzzle loading. It was recognized that the solution might involve a bullet with a diameter smaller than the bore for easy loading. Once in the barrel, the bullet would somehow need to change size or shape to fit the rifling more snugly on its way back out. Some efforts involved fixing odd steel projections at the bottom of the bore of the gun, which were designed to help mash an undersized soft lead bullet into the rifling grooves. More success ultimately came about from changes in projectile designs. Perhaps the earliest movement in this direction was made by British Captain John Norton, who in the 1820s devised a special bullet that, like a musket ball, was of a smaller width than the gun bore for which it was intended. Because of this windage, it could be loaded fairly easily. The real secret to this bullet was a slightly hollowed-out base that was rammed up against the powder charge. When the gun was fired, expanding gases filled this hollow space up and as they began to shove the bullet down the barrel, they also pushed the thin sides of the "hollow" outward into the rifle grooves. Thus the projectile essentially had one small diameter for loading purposes, and a larger one upon firing. In spite of successful testing, the bullet was rejected by the Board of Ordnance perhaps because it was simply too radical a departure from the traditional spherical design (Ruffell 1997).

William Greener presented a similar bullet in 1836. His projectile was oval shaped and had a deep hole drilled almost completely through its center axis. The opening was plugged with a small tapered piece of wood. When loaded, the plug sat atop the powder charge and upon firing was driven into the heart of the bullet, thus pushing its sides outward into the rifle grooves. This ammunition performed exceptionally well but was rejected because the army did not feel comfortable with overly complicated ammunition made up of two separate components. In 1849 a French captain, Claude-Etienne Minie, published the results of his experiments with yet a third type of expanding bullet. This version was "cylindroconoidal" shaped, in other words, a cylinder with a conical top or dome on one end. Although it was not round like Greener's bullet, it did contain a similar rear plug, made of metal in this particular case. Minie also added rings around the outside of his bullet, the gaps between which could be filled with grease to further ease loading, reduce windage, and increase performance. Occasionally these rings would break off within the gun barrel and cause fouling, but on the whole the design was a good one. Despite the fact that the bullet was made in two separate parts, in a momentary episode of sanity the British ordnance board finally accepted this latest version of the expanding bullet and gave Minie £20,000 for his service. Greener was understandably miffed at this inconsistency and suspected that the French officer had stolen his idea. With the support of certain members of Parliament, he was later given a grant of £1,000 in appreciation for what was patronizingly called his "suggestion."

Although a number of variations on the expanding bullet emerged in the midcentury, one last major "modification" appeared. There may have been others who thought of this idea, but credit is often given to James Burton, who worked at a U.S. arsenal. His design was very similar to Minie's except that it had no plug, only a simple hollow cavity at the base of the bullet. This one-piece system was much easier to manufacture and was equally effective. Because Minie's design received more publicity than Burton's, however, the term "minie" was often applied to it. Since "ball" was still the most common expression for a firearm projectile, the bullet was colloquially known as (and mispronounced as) a "mini-ball." The original circular shape of a bullet, however, still lives on today in the continued use of the term "round" to describe a cartridge or a shot.

One thing that did not persist after the development of the mini-ball was spectacular martial fashion. Modern wags sometimes make fun of the colorful apparel of yesteryear's soldiery, for example, pointing out the idiocy of wearing bright red coats in woods creeping with buckskin-clad marksmen. As noted earlier, those situations were relatively infrequent in North American conflicts and downright rare in Europe. In an age where the average firearm was accurate at less than 80 yards, camouflage was largely unnecessary. To the contrary, making an impressive visual display could be a very useful psychological tactic. A thousand brilliantly clothed men marching in precision step across an open field with fixed bayonets was an awe-inspiring, nerve-rattling sight to an enemy.

Fancy duds could also help to pump up the morale of an individual trooper, and uniforms were designed to that end. A large hat or shako made the soldier look and feel bigger. Padded material in the chest and in shoulder pads played a similar role and made him appear stronger. Bright colors, shiny buttons, and fancy epaulets all contributed to a sense of grandeur and self-worth. In the face of inaccurate smoothbore muskets, the morale-boosting effects of these spectacular uniforms more than compensated for the loss of concealment. Military planners and uniform designers were not simply stupid or caught up with fashion. The clothing was functional for the purposes to which it was intended and was well in accord with the dictates of the time. What had started off as a practical means of identifying individual regiments in the early seventeenth century had become something much more important two centuries later. The Napoleonic wars, which saw 2-foot-high bearskin shakos and uniforms with colors as diverse as yellow and pink, represented the epitome of this trend.

The rifle ruined all of this fun. Even in Napoleon's time, riflemen in several states were already wearing dark green, a nod to the dictates of their skirmishing role. Russia, the country that most closely embraced the rifle during this conflict, dressed the majority of its regular musket-armed infantry in green as well. As rifle technology progressed over the next few decades, less exuberant colors continued to come to the fore. The longer ranged mini-ball made its debut in the 1850s, and the decline of uniforms accelerated quickly. The Crimean War, which occurred early in the decade, offered the last real vibrant display of true regimental chic.

THE RIFLED MUSKET

Most troops fighting with the British army in that war were still armed with percussion lock smoothbores, and some were still carrying Brown Bess flintlocks. The rifle units had been issued what was appropriately called a "Minie Rifle." The performance results were mixed, however, one concern being the heavy weight of the gun and its large bore size. A lighter .577-caliber version was designed in 1853, and enormous numbers of the weapon began to be produced at the Royal Small Arms Factory at Enfield. The gun, which informally became known as the "Enfield Rifled-Musket," became the standard firearm of the British army and was a highly successful weapon.

The Enfield differed from earlier rifles in that superficially it looked more like a musket than a Jaeger rifle. It was much longer and was designed to carry a standard bayonet. It was also fitted with a flip-up rear sight that could be adjusted to ranges of up to 1,000 yards. This sounds like optimistic thinking on the part of the rifle's designers, but in one test thirty volunteers firing over a 5-minute time period scored several hits on an artillery-crew silhouette target at 800 yards (Ellacott 1966, 43)! The old title "Queen of the Battlefield" was about to shift from artillery to infantry. As far as enemy foot soldiers went, even moderately skilled shooters armed with Enfields could consistently hit man-sized targets at 200 yards. Cramming a .575 minie bullet down a .577-caliber barrel was still a touch slower than loading a Brown Bess, but the difference in accuracy was now judged worth the minor reduction in rate of fire.

As useful as the minie ball was, it did have one major side effect nobody apparently considered. The British mercantile empire in India had long been dependent upon the use of indigenous Hindu and Muslim soldiers known as "sepoys." The grease used to lubricate the new mini-balls was made from a concoction of beef and pork lard, the consumption of which is frowned upon in each respective religion. Some sepoys became concerned about lubricant from these prelubricated bullets touching their lips after biting a cartridge, and the end result was that a number of them "mutinied" in 1857 against their officers and the British Raj.

The uprising was limited to relatively few units and collapsed in short order, but it did involve horrific atrocities committed by both sides and became a psychologically traumatic episode in the history of the empire. While the Enfield's greasy bullets were not the only cause of the tragedy by any means, the episode certainly demonstrates how catastrophic the consequences of a seemingly minor technological development can be. When John Norton began tinkering around with bullet designs, he may have recognized that some human cost would ultimately be involved. It is doubtful, however, that he would have foreseen the final technology leading to British and Indian subjects being brutally massacred halfway around the world. Ultimately he and his expanding-bullet designer colleagues would tally up a much higher cost.

During the fighting, British soldiers found that they were less easily spotted when their bright red uniforms became dirty. According to legend, some soldiers intentionally dusted up their coats in order to be better camouflaged. The local word for "dust" is "khaki" and was eventually used to describe the light-brown or tan color that the British army came to use in its uniforms. As noted above, with the advent of fast-loading rifled firearms, the psychological importance of a blazing red coat no longer outweighed the benefit of staying hidden.

This point had perhaps become accepted by the early 1860s during the American Civil War. A few units, such as the Franco-Algerian–inspired Zouave Regiments, adopted colorful apparel, but for the most part Federal and Confederate uniforms were not nearly as spectacular as those worn a mere decade earlier in the Crimea. By the end of the first year of fighting, most northern troops had adopted a dark midnight blue coat that disappeared quite nicely in forests and brush. Because southern forces had less regulation to abide by and less access to good dyes, they consequently demonstrated greater diversity in their uniforms. While gray was frequently used, varieties of brown were also common. One ubiquitous dye recipe of nutshells and rusty bits of scrap iron produced a shade loosely known as "butternut" that was very similar to British khaki and made an excellent color for camouflage. If one knew nothing else about the weaponry of the American Civil War, the uniforms alone might suggest serious transformations in the conduct of warfare were underway.

Unfortunately, changes in battlefield tactics were implemented less quickly. For much of the war, soldiers continued to march shoulder to shoulder in neat formations right out in the open. In the age of smoothbore muskets such arrangements made perfect sense, but in the world of the fast-loading rifled musket they were suicidal. Frontal attacks made in this matter were broken up time and time again by deadly long-range rifle fire.

Swords were still employed between opposing cavalry units on occasions, but it was now rare for mounted cavalry to dare take on rifle-armed infantry. More often the descendents of the mounted knight acted in a reconnaissance role and fought dismounted like dragoons. Infantry officers used their swords primarily as badges of rank or as large pointers for directing troops around. Bayonets were likewise of little battlefield application and generally found more useful employment as tent pegs, cooking skewers, and convenient stick-in-the-ground candle holders. The era of stabbing warfare was basically over. Although modern cinema films set in this conflict inevitably include a climatic bayonet fight, encounters of this kind were rare. Out of a sample group of 144,000 battlefield casualties, only 5 percent were actually caused by edged weapons[1] (Dupuy 1980, 171). Field artillery was slightly more lethal at 9 percent, but this means that the vast majority of battlefield casualties were now being caused by firearms.

The actual guns employed during the American Civil War demonstrated a remarkable range of technologies from antiqued weapons of the past to very advanced prototypes of guns that became predominant in the next century. Initially a number of militia and irregular troops were still using smoothbore flintlock muskets. On the other end of the scale, repeating rifles and even some of the earliest versions of machine guns made their debut during the fighting. The majority of soldiers, however, used weapons like the U.S. Model 1861 Rifle-Musket. This gun was based on a line of American muskets that of course were originally related to the French Charleville. After 1803 U.S. armories started to produce some rifles, but not in large numbers. This changed after the advent of Burton's expanding bullet, and Federal facilities began to rebore the barrels of their 1842 model percussion smoothbore muskets with light rifling. In 1855 they produced a new .58-caliber rifle musket with an innovative percussion ignition system that did not prove entirely satisfactory, more about which will be said in Chapter 5. At the outbreak of war in 1861, this basic rifle musket was redesigned with a standard percussion lock resulting in a weapon that was functionally, if not cosmetically, very similar to the Enfield.

This gun, and a slightly modified variant designed in 1863, were both loosely known as "Springfields" during the war and were produced in enormous numbers by Federal and private subcontractors. The quality of these firearms could vary drastically depending upon the integrity of the manufacturer, but the Union was able to build nearly 1.5 million of them

---

1. A popular statistic that is sometimes thrown about claims that there were less than 1,000 casualties of this sort during the entire war. While this figure is a bit too low, deaths caused by cutting-edged weapons were still exceptionally rare.

during the conflict (Davis 1991, 54–58). The Confederacy was eager to capture these guns on the battlefield and their capital's armory manufactured thousands of copies of it, known, unsurprisingly, as the "Richmond."

To meet the demand for firearms, huge numbers of foreign weapons were imported during the war. Many of these performed poorly, but a major exception was Britain's Enfield. The gun was judged to be slightly more accurate than the Springfield and was close enough in caliber size to accept the same mass-produced ammunition. Although the Confederacy ordered far more of these firearms than they were ever capable of actually paying for, around half a million were ultimately imported by both sides. Perhaps the most glaring difference between the Springfield and the Enfield was in the appearance of their locks, barrels, and other metal components. Like the original Charleville, later U.S. muskets and rifled muskets used shiny, unfinished steel designed to sparkle, flash, and awe. The British weapon was also shorter, lighter, and used more brass in its furniture in the tradition of the Brown Bess. The technique of browning too had also long since proven itself, and the Enfield used a dark barrel. In an age of rifling, shiny guns were no more practical than colorful uniforms and would meet a similar fate. By the time the 1861 Springfield was being designed, it was already obsolete in a key function. The minie ball had sped up the rate of fire for muzzle loaders, but other faster reloading firearms had been developed that rendered the rifled musket an antiquity in its youth. Like so many technologies of the industrial and postindustrial age, the sudden obsolescence of relatively new designs would become a fairly common pattern for the firearm in the late nineteenth century. One might conclude that the firearm was in something of a midlife crisis as it constantly reinvented itself. In any event, the remainder of the century would continue to see an incredible surge of alterations and innovations, most of which focused upon rate of fire and were dependent upon a new ignition system.

5

Old Metal Heads: The Self-Contained Metallic Cartridge

◆

PERCUSSION IGNITION

The climate on the northeastern coast of Scotland can be a bit dreary, but locals do what they can to find recreation in it, be it the pursuit of trout in the streams or birdies on the golf green. Other birdies can be pursued too, as long as one has the right equipment. The area is a flyway for migratory birds traveling up and down the coastal waters, and hunters have enjoyed a long tradition of fly shooting in this area. One such hunter was the Reverend Alexander John Forsyth, who practiced the art somewhat poorly around his parish at Belhelvie in Aberdeenshire in the 1790s. He blamed his limited success upon the sport's current state of technology.

In fact a number of improvements in waterfowling arms had occurred over the past century. As noted, previously barrels began to be made of lighter weight early in the century. At one time, an excessively long tube was also needed to keep the pellets from scattering wildly but, more recently, barrels had been shortened through several innovations. One was a technique called "choke boring," whereby the bore was partially narrowed or "choked" at the muzzle. This helped keep shot grouped together longer (Ellacott 1966, 30–31). Henry Nock, a famous gunsmith who created a specially designed breech, further improved the efficiency of the lock itself. In Nock's system the ignition flash was directed into a smaller powder chamber behind the main charge. Rather than igniting on the side and burning inward in an uneven fashion, the propellant ignited directly from the rear. This caused a uniform and efficient deflagration that burned very efficiently. Guns fitted with Nock's system fired faster and with more punch, allowing sporting arms to be fitted with barrels as short as 30 inches (Peterson and Elman 1971, 124). Barrels became light enough that in the late eighteenth century two of them could be welded together, either over and under or side by side, hence the emergence of the famous "double-barreled shotgun."

These weapons still used the flintlock mechanism, of course, and while the ignition system had reached a high level of sophistication, it did not meet Forsyth's exacting demands. Many other hunters were able to shoot birds in midflight with flintlocks, but the self-conscious parson looked for another culprit to explain away his poor marksmanship. When a flintlock fires, there is a slight delay between the time the trigger is pulled and the point at which the gun actually fires. There is also the matter of a small flash in the priming pan momentarily before the main powder charge ignites. According to some accounts Forsyth actually claimed that this little flash startled the birds, which were then able to avoid the oncoming shot. If accurate, this either represents one of sporting history's most pathetic displays of self-excusing, or proves that the waterfowl around Aberdeenshire were once able to move at close to light speed.

When not missing warp-drive ducks, another interest of Forsyth's was the natural sciences and chemistry, subjects he had studied at the University of Aberdeen. A series of earlier French scholars, from Peter Bolduc in the early eighteenth century to Claude Louis Berthollet in the 1780s, had conducted experiments with substances known as fulminates. These can be created by reacting certain metals such as gold, silver, or mercury in nitric acid; allowing the mixture to dry; and collecting the residue. When the resulting material is further treated with pure ethyl alcohol, it crystallizes. The resulting crystals can be crushed (very carefully) into a highly volatile powder. These fulminates, however, are extremely unstable, and while conducting such experiments, Berthollet managed to blow up his laboratory not once, but twice. As they would have said on the comedy series *SCTV*: "He blowed up twice! He blowed up twice good! He blowed up twice *real* good." Building upon earlier French efforts, in 1799 an Englishman named Edward Howard developed a method of making fulminate from mercury that he tried to use as a substitute for gunpowder. Like Berthollet, he found the chemical too dangerous for general use, but added it to a saltpeter mixture in order to create a special powder for priming purposes. Howard remained very skeptical about whether mercury fulminate on its own could be used to ignite gunpowder.

Forsyth, however, thought this was worth further investigation; and, like others before him, he too managed to blow up his workshop in the process. "He blowed up good too!" This minor inconvenience did not slow him down much, however, and he continued his research. He was able to stabilize the fulminates a bit by mixing them with charcoal, sulphur, and other materials. While experiments with these concoctions in a flintlock flashpan were unsatisfactory, he found that striking a minute amount directly with a hammer regularly triggered a small but intense explosion. Although the mechanisms involved have been redesigned many times over, the basic system of ignition by striking remains in use in most firearms today. The Scottish inventor had figured out that fulminate could be used to ignite gunpowder, but he also realized it would require an entirely different type of gunlock in order to be so employed.

What Forsyth came up with is the percussion lock, named after its manner of ignition. Forsyth replaced the flashpan and battery with a metal container that looked like a small perfume or "scent" bottle. This magazine held a reserve of powdered mercury fulminate and could be rotated into an upside-down position. When done so, a minute amount of primer fell into a smaller chamber, which also housed one end of a plunger. When rotated back, the end was now in an upright position facing a simple flat steel hammer that had replaced the flintlock's cock. When the trigger was pulled, the hammer slammed down on the plunger, which in turn struck the mercury fulminate charge. A small jet of fire from the miniature explosion was funneled through a closed touchhole into the barrel, where it ignited the main charge. Forsyth was apparently so content with his lock's basic design that he chose to never significantly modify or improve on it in later years. There were plenty of other inventors, however, who were willing to try and work around the reverend's existing patents. By 1820 a number of inventors and gunsmiths, including Joseph Manton, had experimented with new percussion designs.

This work eventually led to something called the percussion cap, and it is not clear who deserved final credit for the invention, although Joshua Shaw of Philadelphia was a likely contender. A small touch of mercury fulminate compound was set inside a metal cup shaped roughly like a top hat. Initially these iron caps were intended to be reusable, but ultimately they came to stamped out of cheap throwaway copper. In a typical percussion lock, the cap was slipped over the top of a small cylindrical "nipple" that had a tiny hole drilled through it lengthwise. Unlike Forsyth's system, there was no priming magazine; instead the nipple's hollow center led directly into the closed touchhole. When the hammer snapped down on the top of the cap, the fulminate exploded, and with nowhere

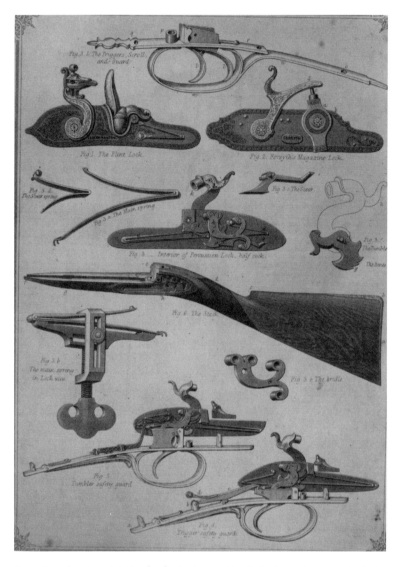

Illustration showing details of a flintlock, scent-bottle lock, and percussion lock.
© Hulton Archive/Getty Images.

else to go, the flame burned down the touchhole into the main powder charge.

This second system of percussion ignition became the predominate one and Forsyth's original disappeared. However, the fulminate scent-bottle magazine may have been a superior idea. Digging around in one's pocket or "cap box" for a cap to be placed separately on the nipple is not always quick or convenient. With Forsyth's system, a mere twist primed the weapon minus the fumbling. While the scent bottle was efficient, it did require precision manufacturing and was thus expensive to produce. The percussion cap was a more practical design for mass production.

After three centuries of flint-ignition gunlocks, there was bound to be some resistance to the new system. Critics claimed that percussion-equipped guns did not fire with the same power as a flintlock. In fact the opposite case was true, as none of the force of the blast was lost shooting back out of an open touchhole. A much more serious critique was the fact that the caps would sometimes explode, hazarding the shooter's eyes with a shower of metal fragments. This was partially resolved by reducing the amount of fulminate in the cap to a bare minimum and by adding metal shielding, either around the nipple or extending down from the hammer. This problem was never entirely solved, however, and the most advanced versions of percussion locks found on Springfield and Enfield rifle muskets would occasionally send a jagged cap fragment tearing across the shooter's face. A number of modern-day American Civil War reenactors who use versions of these guns sport telltale jagged scars on their cheeks.

Against these criticisms, the percussion lock offered significant advantages to the shooter. The system was far less prone to misfire than was the flintlock. Being almost entirely enclosed, it was also much more resistant to foul weather. This had particular appeal to sportsmen, and it was perhaps no coincidence that the origins of the lock lay with a frustrated duck hunter. It was amongst private sportsmen in particular that the new design attracted the most interest, and hunting guns with percussion locks quickly became standard in the 1820s. In many cases, old flintlocks were simply refitted with a new mechanism.

As might be expected, however, the tradition-oriented military of the nineteenth century was a little less enthusiastic about the innovative system. This was not entirely due to a pig-headed stubbornness against change, but rather with concern over the rate of fire. The flintlock could be conveniently primed with a little powder from the cartridge, but the percussion lock required that additional step of retrieving a separate cap. Undoubtedly soldiers in action could circumvent the problem to some extent by keeping a small store of caps held in one palm, although this too is a touch cumbersome and requires some sleight-of-hand skill. What probably won over military officials was the realization that there is a difference between theoretical and *effective* rate of fire. While finely made flintlock dueling pistols rarely misfired, more crudely constructed military locks frequently did. An individual soldier could reload a flintlock faster than a percussion weapon, but that fact does not consider the number of times the older lock failed to actually discharge. If one multiplies that weapon's rate of misfire by hundreds of soldiers firing hundreds of shots, the extra step involved in the percussion system becomes less of a problem. Sometimes bad weather could be catastrophic to formations equipped with flintlocks; one of the few reversals the British infantry suffered during the Opium Wars with China occurred during foul weather. Gradually percussion locks began to appear on military weapons. As noted previously, the Brunswick was the first British military long arm so equipped, which was perhaps the only advantage that notorious gun enjoyed. In 1838 Britain decided to adopt the lock for all new military firearms. France followed suit in 1840, and the United States did likewise two years later (North and Hogg 1977, 83).

Some military officials still worried about the rate-of-fire issue, and the United States took an unusual step to address this concern. A copper cap is not the only way to achieve percussion ignition, as a New York dentist with an interest in firearms discovered. Dr. Edward Maynard realized that paper, which was even cheaper than copper, could perform the same function. In 1845 he took a thin strip of paper and arranged tiny pellets of priming material along it. A second strip of paper was glued on top, creating a "tape" of primer. Using industrial manufacturing techniques, miles of this stuff could be cranked out cheaply, and Maynard set about developing a firearm lock capable of using it. The tape was coiled like a snake and placed in a circular magazine set partially into the side of the lock itself. When the hammer was cocked, an arm pushed a piece of the tape forward, placing a priming charge over the nipple. This hammer not only detonated the fulminate, but it also sliced off the used section of tape. The Maynard lock was similar in some ways to Forsyth's original scent bottle in that it used the percussion system, but with a handy magazine of full of primer. It eliminated the extra step needed to place a cap, and was even easier to prime than a flintlock. Also the fulminate in a Maynard lock could not all go off at once, a frightening possibility with Forsyth's original scent-bottle system. The U.S. Ordnance Board was hesitant about the design but the secretary of war, future Confederate President Jefferson Davis, was so enthusiastic that it was adopted for use on the U.S. Model 1855 rifle musket (Hicks 1957, 82–84).

Again, field performance did not quite match the sanitary conditions of the testing ground, and the weapon soon began to develop a negative reputation further down the ranks. In spite of efforts to waterproof the tape primer, moisture proved to be one of the biggest problems. The paper strips were susceptible to adverse weather, even humidity. The Ordnance Board eventually recommended a return to the older separate percussion cap

(Gluckman 1948, 229). This advice was acted upon, and the Model 1861 Springfield was developed using the slower but more dependable system. The Maynard priming tape remains in service in toy cap pistols today, although their performance continues to be spotty. In a case of history repeating itself, the 1970s saw the introduction of plastic percussion-cap toy pistols that had a much higher level of reliability, albeit at a higher cost per shot.

REVOLVERS

More promising efforts to improve the rate of fire through the use of percussion ignition were well underway by the time the Maynard system was adopted. As noted previously, firing as many shots as possible in a short period of time has been an ancient concern of humans. Since the advent of the gun, a number of ingenious efforts had been made to address it. One of the more inane solutions was to equip a single barrel with more than one firing lock arranged in succession along it. The barrel was then charged with multiple shots, each hopefully aligned with a lock. Theoretically the shooter would pull the trigger on one lock at a time, the foremost one first. The potential for accident in such a system is an insurance underwriter's nightmare. An excited person in the heat of battle, or in the thrill of the hunt, might accidentally fire a rear charge first. Furthermore, even if fired in proper succession, sparks from one shot might set off other charges. In spite of the obvious potential for harm, multilock single-barrel weapons were being constructed as early as the sixteenth century. Examples of these guns vary widely, from a three-charge German wheel lock pistol of the 1560s to a four-charge English flintlock pistol of the 1770s to an Italian two-charge percussion hunting rifle built in 1837.

An even stranger, or more stupid, variation on this technique reportedly involved firing multiple charges out of a single barrel by using a single lock. The suicidal gunner had to carefully prepare his ammunition prior to shooting by drilling a hole through each lead ball. A lock fixed far forward on the barrel ignited the first charge, part of the flame from which passed through the hole drilled in the second ball and ignited this charge. The whole mechanism was supposed to work like a roman candle, with bullet after bullet launching from a single gun until the charges were exhausted. Some of these guns could fire sixteen shots in this manner (Blackmore 1964, 42–43). This system certainly lost any element of surprise after the first discharge, and bullets with holes drilled through the center could not have been remotely aerodynamic. Cramming a stack of powder and charges into one barrel inevitably risked disaster. A particularly mindless version of this technique built in Germany around 1660 combined multiple loads set in multiple *barrels*. This device was made to fire a total of twenty-nine shots with one pull of the trigger (Blackmore 1964, 42–43; Reid 1984, 164–165). One shudders to imagine the effects of a misfire somewhere along the way.

Increasing the number of barrels attached to a gun but loading each barrel with a single charge was a much safer method. Late eighteenth-century shotguns were not the first weapons to be so fitted by any means. One of the oldest surviving examples of a medieval handcannon, dating perhaps from the late fourteenth century, contains four barrels fixed to a single tiller. A watercolor illustration of Emperor Maximilian I's arsenal, painted about 1505, shows both double- and triple-barreled matchlocks. A number of later flintlock and percussion pistols carried anywhere from three to seven barrels either stacked together in a boxlike pattern or spread out horizontally in a splayed design known as the "duck's foot." This latter supposedly was intended for crowd control. In some of these weapons, a single pull of the trigger discharged all the barrels simultaneously or in quick succession. One of the most extreme versions of this approach was a nightmarish flintlock carbine developed in part by Henry Nock that fired seven barrels at once. Pulling the trigger on such a monstrosity had to have been an unnerving experience. This weapon really shares more in common with the shotgun or blunderbuss and was not intended to improve rate of fire.

There were more refined and subtle versions of the multibarreled firearm. For example, wheel lock pistols sometimes were built with two barrels set in an over-and-under style, but with two separate locks and triggers fixed on each side. The gun had a perfectly straight handgrip, and after one shot the weapon was simply turned upside down in order to fire again. One variation on the multiple-barrel approach utilized two barrels that could be rotated in place over and under each other. These guns had one trigger, but each barrel was equipped with a separate flashpan and cock, allowing them to be preprimed and fired in quick succession. The system even came to be employed on fine English flintlock pistols and Pennsylvania rifles, although the extra weight must have been more inconvenient in the second case. A similar technique was to use a single lock to fire three or more rotating barrels. Since the barrels revolved around each other, this type of weapon is known as a revolver.

Like so many other modern "innovations," the revolver was actually a much older concept, predating Samuel Colt by centuries. One Venetian matchlock version with three rotating barrels has been dated to the 1540s. In later centuries, similar flintlock variations were experimented with. These early revolvers had to be turned by hand and reprimed after each shot. Because the percussion cap allowed the ignition mechanism to be simplified, however, more efficient and effective versions of this gun could be constructed in the early nineteenth century. Several of these "pepperpot" or "pepperbox" pistols emerged in Britain toward the end of the Napoleonic wars. Named after their apparent similarity to spice holders, these revolvers typically carried four to six barrels, each set with its own percussion nipple. The barrels were initially rotated by hand, although later versions revolved automatically when the hammer was cocked. Like earlier multibarreled pistols, however, these revolvers tended to be fairly end-heavy and unbalanced.

A really superior weapon would need to have the weight of the single-barreled gun combined with the rate of fire found in a multibarreled weapon. The solution involved using just one barrel, but multiple breech chambers. In essence the revolving barrels were reduced to their shortest possible length, just long enough to hold powder and shot. One long barrel was then fixed in place directly ahead of, and abutting against, this rotating cluster of mini-barrels now called the cylinder. The chambers of this cylinder, each holding a loaded shot, could be revolved in turn to match up with the fixed barrel. In this way, it was possible to fire a number of charges from the separate cylinder chambers through the single barrel. Such a system greatly improved the balance of the gun and essentially represents the common version of the revolver that is still in widespread use today.

Like so many other important developments in the history of guns, the first prototypes appeared far back in the early childhood of the firearm. One arquebus-like weapon with a revolving chamber was built by Hans Stopler of Nuremberg in 1597. It used a snaphaunce lock (snaplock) and was fairly advanced in design, suggesting that earlier models may have once existed. In addition to this arquebus revolver, there were even more familiar-looking pistol revolvers. The problem with these weapons, like most early modern breechloaders in general, was the loss of gas from ill-fitting parts. Loose machinery work could also allow stray sparks from the chamber being fired to find their way into other loaded chambers with disastrous results. It would take the precision manufacturing of the early industrial age to make this design workable.

By the early nineteenth century, one gunsmith had achieved sufficient skill to produce a reliable flintlock revolver. After failing to attract local interest in his design, Elisha Collier of Boston traveled to London and took out a patent on a specially designed revolver in 1818. Collier's revolvers were extremely well made and quite safe to operate. They contained an intricate self-priming mechanism that automatically charged the flashpan as the cylinder was rotated. Most of these revolvers were fitted to pistols, but a few found their way onto rifles and fowling guns. Well-constructed as the weapons were, unfortunately for Collier their flintlock ignition systems were already going out of use by the time he developed his design. Still, a number were purchased for colonial service in India.

The percussion system offered new opportunities for the revolver that were seized upon by another New Englander, Samuel Colt. Colt's particular application was so successful that countless later enthusiasts credited the gunsmith with actually inventing the revolver. Legends factor heavily in the story of the Colt pistol, and one of the more famous yarns has young Samuel carving revolver components out of wood while serving aboard the aptly named *Cotlo* in 1830 at age sixteen. It is more likely that he saw one of Collier's pistols while in Calcutta in 1831 (Edwards, 1953, 22–23). Upon his return he began experimenting with metal prototypes, the cost of which was funded by popular "scientific" lectures on laughing gas by "Dr. Coult." By 1836 Colt had patented his basic design in both the United States and the United Kingdom, and set up a manufacturing company in Paterson, New Jersey.

Colt's basic system simply fixed a percussion nipple onto a slight recess at the rear of each individual cylinder chamber. The enclosed nature of percussion-cap ignition along with the recessed nipples greatly reduced the chance of other cylinders accidentally discharging. He also added a mechanism that rotated the cylinder automatically and lined up a new chamber each time the hammer was cocked. Colt's was not the first revolver to incorporate this feature, but it was just one element in the successful package he managed to pull together. His first model was a long-barreled, light-caliber pistol known as the Paterson, which also utilized an interesting hidden trigger that popped out of the handle when the gun was cocked. Colt's initial venture had failed by 1842, but his pistols won praise in service against the Seminole and Comanche nations. A timely war with Mexico generated further demand for revolvers, and with some new financial backers Colt began winning government contracts. He set up business again in Hartford, Connecticut, and even in London for a few years.

The new pistols Colt produced from the late 1840s through the 1860s kept the standard design, but included a few refinements. The barrels were shortened, sported fixed triggers, and usually were of heavier calibers. They also included a clever lever-driven plunger attached under the barrel that replaced the function of the ramrod. The pistol chambers were loaded individually through the front along the side of the cylinder, initially as loose powder and shot, and later in premade paper cartridges. Each charged chamber in turn could then be rotated to align with the plunger and be packed more tightly with a simple pull of the lever.

Colt experimented with a variety of these revolvers, but the most famous were the so-called Army and Navy models. The Navy emerged in 1851 and was not really designed as a naval weapon per se, but probably picked up its name informally from a nautical scene inscrolled on the side of the pistol. Although the gun was well made, it used a .36-caliber bullet that many found too light to be practical. The most successful of the larger bore pistols was the .44-caliber 1860 Army model. Once again Colt's luck was perfect and the gun appeared just in time to see wide service in the American Civil War. The fact that thousands were used in this conflict points to another key aspect of Colt's expertise: the ability of his company to mass-manufacture firearms. Of course he did not quite develop the assembly line—this would be left to Henry Ford—but Colt's facilities used the most modern industrial equipment and production techniques available at the time. He achieved an exceptionally high degree of standardization in firearm parts, something inventor Eli Whitney had earlier championed but never really achieved. Colt's standards were so precise and his firearms were so uniform that it was possible to carry several extra preloaded cylinders that could then be slapped into the frame of the gun for quick reloading.

A Colt-style cap and ball pistol immortalized in a Confederate monument at the Gettysburg battlefield. Notice the soldier toward the rear biting a paper cartridge for use in his rifled musket. © 2002 N. Carter/North Wind Picture Archives.

Although the most common application of the revolver system was found in pistols, there was nothing to prevent the mechanism from being applied to longer firearms. A number of Colt pistols themselves, particularly the large "Dragoon" models, could be fitted with a detachable buttstock and used as a rudimentary carbine. The company also manufactured revolving rifles in a hearty musket-like .56 caliber. One would think these guns would have been as popular with the military as was the revolving pistol, but the cost of such advanced weaponry for common infantry probably weighed against their widespread use. Furthermore, the Colt pistol's reliable character did not extend to the rifle, which had a reputation for exploding.

THE BREECH-LOADING RIFLE

If nineteenth-century gunsmiths had now become capable of mass-producing gastight revolvers, they certainly could revisit other old firearm experiments of the past. One of the more successful efforts toward this end was the revival of the breechloader. This approach, actually, had been around since the late Middle Ages and although Ferguson's particular version never really took off, others would apply similar methods to rifles, pistols, and shotguns. Providing that one could construct a breechloader that did not leak excessive gas and/or explode, the potential advantages of the system are obvious. By eliminating ramming, a major cumbersome step in the reloading process was alleviated. Removing a ramrod from its holding place under the barrel, inserting it into the muzzle, repeatedly hammering down the charge, and returning the device to its proper place involves a lot of effort and a high degree of fumbling around. Not only was the breechloader faster to charge, but it was also much easier to do so while prone or on the move. A host of different designs attempted to tackle the challenge of breech loading, although not all met with equal success.

One of the more notable post-Ferguson attempts to create a breech-loading military rifle was undertaken by John Hancock Hall of Maine. He later claimed to have had no knowledge of the Ferguson weapon and was primarily motivated to improve the performance of rifles so that success in combat would be less dependent upon the formations and drills that professional European soldiers excelled at (Huntington 1972, 2–3). Hall patented his particular version of the breechloader in 1811 and began producing prototypes later in the decade. His initial models used a flintlock ignition system, but he switched to percussion in 1833. All of his guns, smoothbore or rifled, long-arm or carbine, flintlock or percussion, used a similar rear-loading mechanism. A release latch near the trigger guard allowed a hinged breechblock to swing upwards out of alignment with the rest of the barrel. This exposed section, which actually pointed in the direction of the muzzle, could then be charged with powder and shot before being snapped back down in place. Although the basic idea was sound, and Hall recognized the importance of durability, the design was constantly modified, adjusted, and tinkered with over the next three decades. The release latch snagged on clothing and branches, and the weapons suffered from a high degree of gas leakage, particular after seeing a little wear and tear. In several cases individuals simple gave up on the gun as a rifle, removed the breech mechanism, and used this section as a crude pistol. Although the gun won wide disfavor during the Mexican War, the basic theory was sound enough to inspire the Norwegian military to develop a similar pop-up rear-loading carbine in 1842. Hall also managed to achieve the full interchangeability of his weapons' component parts even before Colt did. This, and the advanced production techniques he pioneered, led to a type of manufacturing known amongst Europeans as "the American System" (Smith 1977, 219–251). In this sense he may have been more of an industrial pioneer than Henry Ford, even if Ford's cars actually worked reasonably well and Hall's guns did not.

A more successful design that allowed the breechblock to slide backwards and forwards by means of a pop-up handle was created by William Jenks of Columbia, South Carolina. When the handle was pulled up and backwards, the sliding breechblock exposed a small hole in the top of the barrel. The bullet and powder were loaded through this opening, which was then closed by sliding the breechblock forward. This sliding breech was locked firmly in place by a separate side spring. Jenks began producing rifles and carbines fitted with his lock in the late 1830s and although tests proved that the weapons were remarkably effective and durable, they received a cool reception from the military. This may have been a result of general suspicion about breechloaders in the wake of Hall's troublesome firearms (Butler 1971, 143–144).

A certain degree of suspicion initially met another well-made breechloader patented in 1848 by Christian Sharps, who had worked with Hall in the 1830s and thus understood the challenges of breechloaders better then most. Like the much-earlier Ferguson rifle, this lock cleverly made use of the trigger guard to open the breech. When pushed down and forward, the trigger guard lever pulled a rectangular section of the breechblock straight down and exposed the rear opening of the barrel. A cartridge could then be slipped in, and as the trigger guard was pulled back into place the breech was closed. The breechblock also had a cutting edge on it that sliced the rear end of the cartridge off, exposing loose powder to the interior of the touchhole. This last feature was somewhat redundant as a number of weapons, including

Colt's revolvers, began to employ cartridges made of paper or linen treated with nitrate, similar to matchcord. Not only did the cartridge itself ignite easily in order to set off the charge, but its remains would also burn up more completely in the barrel. Sharps's guns could be primed by a standard percussion cap, although one 1855 version used the Maynard tape system. A number of Sharps's models incorporated a clever automatic percussion device invented by an associate, Richard Lawrence. This system used fulminate pellets set in a spring-loaded magazine that popped into place as the gun was cocked and fired.

Sharps's initial weapons continued to suffer some problems with gas leakage, but these were largely addressed after 1853 by a special breech seal. This consisted of a metal ring about the diameter of the barrel, which was recessed into the breechblock and could move just slightly backward and forward. When the gun fired, gas from the charge itself was forced under the edges of the ring and drove it against the rear of the barrel. The force of the expanding gas essentially helped seal the force of the expanding gas. Sharps fitted this lock onto rifles and even a shotgun, but his design became most widely employed on cavalry carbines. Why this was the case is not entirely clear. The potential advantage of a breechloader to all armed forces should have been obvious. A report from the U.S. secretary of war in 1859 stated, "With the best breech-loading arm, one skillful man would be equal to two, probably three armed with the ordinary muzzle-loading guns" (Fuller 1965, 225). Nevertheless this apparent fact did not create appropriate action, and most infantry continued to use older styled weapons. Perhaps the need for cavalry to be able to reload easily on horseback was considered more pressing.

The Sharps Carbine would see action during the American Civil War, but it was not alone. There was a great variety of breech-loading cavalry carbines on the scene made by both northern and southern gunsmiths. Some of these were quite innovative in one fashion or another. As an example, Ambrose Burnsides, arguably the most incompetent high-ranking American military commander in history, found success off the battlefield with a breech-loading carbine that he designed prior to the onset of hostilities. His model used an upswinging breechblock, much like the earlier Hall's rifle. Unlike it or the Sharps breechloaders, however, Burnside's gun used an odd-looking type of metal ammunition with a small hole in the center near the cartridge. The charge was ignited by a regular, separate percussion cap that simply sent flame through the hole. The breech was sealed by a usual-looking brass ringlike projection that jutted out where the .54-caliber mini-ball was seated in the brass tube. The cartridge looked a bit like a pointed scoop of ice cream in a flat-bottomed brass cone. The system performed

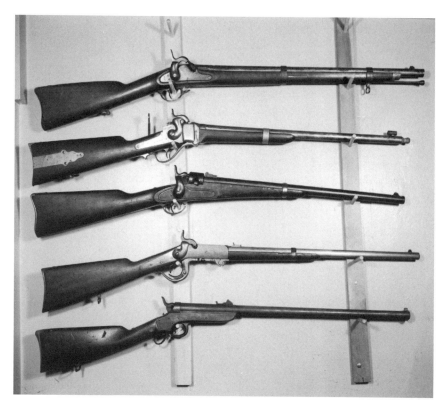

American Civil War carbines. The top weapon is a typical percussion muzzle-loading weapon. The second from the top is a breech-loading Sharps, and the second from the bottom is a breech-loading Burnside. © Hulton Archive/Getty Images.

well, though, and Burnsides's gun earned a distinguished military reputation, even if its inventor did not. The basic Sharps system remained predominant, however, and it has been estimated that over 100,000 of them saw action.

While the majority of the Sharps guns used in the war were cavalry weapons, noteworthy service from the infantry versions came in the hands of Colonel Hiram Berdan's U.S. Sharpshooters. Berdan, a noted marksman and gun tinker, raised two regiments of special troops for service early in the war. The requirements were stringent: volunteers reportedly had to fire ten consecutive shots into a 10-inch circle at 200 yards from a standing position. They were initially given Colt's revolving rifles, but according to tradition their shooting skills so impressed President Abraham Lincoln that he personally intervened to see that they were issued the superior Sharps infantry rifles. Like earlier rifle units, Berdan's men were dressed in green and trained to fight in small units as skirmishers. They also took advantage of rough terrain and avoided exposing themselves to direct fire. Despite the efforts at concealment, these units took heavy losses of around 40 percent throughout the course of the war. This statistic, however, was a result of the excessive amount of action the sharpshooters saw. The reputation and effectiveness of these regiments were such that Federal commanders came to rely upon the men in green and employed them constantly in skirmish and scouting operations. Their accurate, rapid-firing Sharps rifles were certainly a large part of the formula for success. They could deliver such an enormous volume of accurate fire that Confederates who encountered them frequently overestimated their actual numbers. It must be noted, however, that other quick-firing gun designs were on the scene by the time of the American Civil War that would make the Sharps rifle seem positive sluggish in comparison.

Just as breechloaders drew inspiration from older designs, these very advanced weapons too were the end products of much earlier innovations in firearms technology. To understand the next stage in firearm development, one must again step backwards to find the earliest innovations leading up to it. In this particular case, the setting was the Napoleonic period and the inventor was a pretentious Swiss gunsmith with a somewhat uninspiring last name. Born to a wagon-maker in 1766, Samuel Johannes Pauly served as a sergeant major in the artillery before making his way to Paris in 1802. Once there, in the manner of "Dr. Coult," he adopted the more impressive title "Colonel Jean Samuel Pauly" and began tinkering with firearms. Pauly was impressed by Forsyth's scent-bottle lock and sought ways to improve upon it. He eventually produced a radically different firing mechanism, but before doing so, he made an important innovation in ammunition. He recognized that the fulminate itself did not have to rest in a separate magazine, percussion cap, or paper strip but rather could be fixed to the actual cartridge itself. Pauly designed several types of ammunition utilizing both full-brass casings and partial half-casings fitted to the rear of a larger paper cartridge. A small bit of fulminate was fixed in a miniature pan on the center back of the metallic casing, which is known ironically enough as the "head" of the cartridge. The fulminate was ignited by means of an internal metal spring-driven "firing pin" instead of a side-mounted percussion hammer. Once the charge in one of these "center-fire" cartridges ignited, its metal sides were driven outwards toward the bore, thus helping to "obturate," or close off, the gases. The rear of the cartridge in effect acted somewhat like the breech seal found in the later Sharps.

Pauly played around with several lock designs for the use of these cartridges, but the most promising was something now known to us as the hinge action. In this system, the gun almost splits into two sections held together by a hinging mechanism. The barrel and front part of the stock (known as a "fore stock") swung downwards away from the lock, thus exposing the open breech of the barrel. A cartridge was simply inserted into the rear, and the front part of the gun was snapped back into place, ready to fire. Pauly patented this system in Paris in 1812 and made an impressive demonstration for military authorities, reportedly firing twenty-two shots in 2 minutes. Napoleon declined to pursue a contract, perhaps fearing that the mechanism was too delicate for field service or that the cost of metal cartridges was too high. Imagine if he had several regiments so well armed at Waterloo! In any event he did not, and after the emperor's forced career change, Pauly traveled to the "enemy's" camp, where he altered his name again to the Anglo-Saxon–sounding "Samuel John Pauly" and partnered with Durs Egg, a master London gunsmith. The two experimented with further designs, including a novel compressed-air ignition lock, but with little real financial success. The odd thing about this whole episode is how advanced Pauly's system was. Although his weapons never gained popularity in his lifetime, eventually the metallic self-primed cartridge would become an almost universal feature of later firearms. In addition to this pioneering ammunition, the hinge-action breechloader design became a common feature in many sporting arms. One of Pauly's former assistants, Casimir Lefaucheaux, produced a hinge-action double-barreled shotgun in 1836 that was exceptionally successful, and some of the highest quality shotguns made in the world today still use this basic design.

The center-fire cartridge, on the other hand, had to wait decades longer before being rediscovered. While Pauly's guns enjoyed little popularity in his lifetime, the same was not true for those of one of his most astute assistances, Johann Nikolaus von Dreyse. After working in Paris, von Dreyse traveled back to his Prussian homeland and began manufacturing percussion caps. He came to the conclusion that the ideal place to seat a percussion cap would be inside the powder charge itself. Von Dreyse designed a special cartridge in which the fulminate was set at the base of the bullet, in front of (and surrounded by) the powder charge. To strike this buried ignition material, it was necessary to pierce a relatively long way through the rear of the paper cartridge. Von Dreyse therefore changed Pauly's firing pin into a much longer and thinner firing *needle*. Instead of employing the hinge-action mechanism, however, von Dreyse fixed the needle to a sliding breech-block similar to that found in Jenk's breechloader. This particular lock, however, had an extended knob that jutted out from the top of the breech-block. The bore could be exposed by grasping this knob, shoving it to the shooter's left, then pulling it backwards. This action also cocked an internal firing pin. After the cartridge was slid into the rear of the barrel, the knob was pushed forward and to the right, locking the breechblock into place. The whole operation is similar to the opening and closing of a door bolt, and the name "bolt-action lock" came to be employed to describe this sort of mechanism. Von Dreyse's particular version was more noted for its unusual firing pin, and the term "needle gun" came to be used to describe it. This 1838 invention so impressed the Prussian military that they began rearming their troops with it in the following decade. This was a farsighted decision as the firearm was not without fault. The long, thin needle was a tad fragile and inclined to breaking, particularly after some use had exposed it to the corrosive effects of gunpowder. Like the Hall breechloader, the gun also suffered from a fair amount of gas leakage at the breech, and some critiques questioned the weapon's accuracy in comparison to that of other contemporary rifles. The needle gun also lacked the range of an Enfield or Springfield, but as later German military officials would eventually recognize, relatively little combat was every really conducted at 800 yards. Von Dreyse's weapon, on the other hand, could fire as many as seven shots a minute, and by adopting an advanced breechloader which such a fast rate of fire, Prussia was well ahead of its neighbors in the armament of its basic line infantry (Ross 1979, 175–176). This point was driven home in victorious wars with Denmark in 1864 and Austria in 1866, after which every major European state scrambled to obtain breechloaders. By this time, of course, there were many highly functional variations to choose from.

For example, France had adopted a superior version of the breechloader by the time Prussia went to war with it in 1870. The design was basically a modified version of the needle gun itself, made by Antoine Alphonse Chassepot of the Chatellerault arsenal. Instead of having the firing pin pierce into the cartridge itself, it used a center-fire system similar to Pauly's earlier design. This avoided the effects of gunpowder corrosion on the firing pin. He also attached a ring made of rubber to the sliding bolt in order to seal the breech. Although these would wear out after some use, the basic success of this device gave the Chassepot's smaller .43-inch (or 11mm) caliber bullet a higher muzzle velocity. Although France ultimately lost the Franco-Prussian War of 1870–1871, its basic infantry service arm was generally reckoned to be more accurate and longer ranged than the victor's needle guns.

THE REPEATING RIFLE

As witty as these devices were, they really do not represent a fundamental improvement over the self-contained percussion-cartridge systems of Pauly and Lefaucheaux. A much more successful movement in that direction came in a long and roundabout way starting with the efforts of Walter Hunt back in the 1820s. Like Pauly, Hunt too was a clever inventor with little business sense who failed to reap much that he had sown. He reportedly invented the safety pin but sold the rights for $100 to pay off a debt, and also never quite got around to patenting his prototype "sewing machine." Hunt also took a stab at firearm inventions and developed a unique ammunition design. The bullet itself was hollowed out, much like a minie ball, but the empty space was filled with powder. This end was sealed by a metallic skirt with an open hole in the rear. Hunt patented this "Rocket Ball" ammunition in 1848 and then built a firearm designed to use it called the "Volitional Repeater."

Like so many other weapons of the nineteenth century, his gun was similar to a much older, but never fully perfected, design. A seventeenth-century Danish marvel known as the Kalthoff gun used a cylindrical tube mounted under the barrel that held multiple balls. This remarkable wheel lock also had a powder magazine in a hollowed section of the stock butt, and the whole gun could be primed and loaded by pivoting the trigger guard. As brilliant as the design was, it was expensive and reportedly a bit delicate. Hunt's elaborate and likewise frail Repeater, later renamed "Volcanic," used some similar elements. It employed a tubular spring-loaded bullet magazine under the barrel that fed ammunition into the breech chamber with a finger lever.

Over the course of the next decade, the design and accompanying patents then bounced through a byzantine series of buyouts, improvements, and business transactions. At one point the evolving repeater passed through the hands of two inventers, Horace Smith and Daniel Wesson, who modified the Rocket Ball cartridge by adding a bit of fulminate backed by a metal disk to the rear. The new design removed the need for a separate percussion cap, and eliminated an extra loading step. The firearm was about to take another major leap forward in rate of fire.

In the short term, though, like many earlier breechloaders the modified basic Volcanic Repeater gun was plagued by gas leakage, and its unique ammunition only contained a very small powder charge. The end result was that the Volcanic Repeater was not all "volcanic" and suffered from an exceptionally low muzzle velocity. At the moment a Burnsides or Sharps bullet left the end of its respective carbine, it was traveling anywhere from about 950 to 1000 feet per second. It has been estimated that the Volcanic Repeater only managed about half of this speed (Butler 1971, 221). Still, the overall concept behind the weapon was promising, assuming the machining was improved and a more powerful cartridge could be perfected.

Smith and Wesson left the Volcanic Repeater Company in 1857 in order to pursue their "side" interest in pistols, but the work was continued under the new ownership of a shirt-maker named Oliver Winchester, who changed the name of the firm to the New Haven Arms Company. As part of the purchase deal, a very interesting clause was included that allowed the Winchester firm to use any further innovations in ammunition that Smith and Wesson might come up with later on. As part of their work on revolvers, the prolific gunmakers continued to experiment with ammunition. One idea they patented in 1860 was to put fulminate in the outer rear edge of the cartridge. The primer circled the entire rim and could be ignited by being struck anywhere along the circumference. This rimfire cartridge was initially thought to be a better ammunition type to build a revolver around because in this type of weapon the striking edge of the hammer was traditionally located physically closer to the edge of the cartridge than to its center-rear. The basic revolver design would not require extensive modifications in order to use this particular type of metallic cartridge. A center-fire cartridge would have needed a separate firing-pin mechanism to discharge, and while later revolvers were constructed in this manner, Smith and Wesson initially decided to simply alter the cartridge. One potential problem was getting the fulminate to spread equally around the rim and stay there; failure to do so might result in misfires. Smith and Wesson addressed this issue with a special plug that helped hold the fulminate in place.

In accord with the arrangement noted above, at the same time that Smith and Wesson were working on their cartridge, Winchester's manager, Benjamin Tyler Henry, was developing similar ammunition for the Volcanic design. His new .44-caliber rimfire cartridge had a relatively long metal jacket that allowed for much greater powder capacity than the older Rocket Ball–type designs had. Henry modified the Volcanic pistol design to create a repeating rifle for this ammunition. A trigger lever was employed that could be worked by the entire hand, rather than one finger, which eased the reloading process. Because it was a rifle, it was able to sport a much longer tube magazine that could hold *fifteen* of the petite Henry cartridges. The bullet was admittedly much smaller than that used in other carbines, but when fired from the modified Volcanic weapon Henry built for it, the round had a muzzle velocity of 1,125 feet per second. This high speed gave the small projectile a harder impact similar to larger and slower bullets. More on the relationship between speed and size of a projectile will be reviewed in the next chapter.

Not only did the Henry rifle's bullet move relatively quickly, but when compared to other contemporary firearms, the gun had an unheard-of rate of fire. By simply shoving the trigger-guard lever down and forward, a spent casing from the gun's barrel was ejected through the top of the breech, a fresh cartridge slid into a lifting mechanism, and the hammer was cocked back in a firing position. Snapping the lever back into place pushed the new cartridge up and into the breech, ready to be fired. This lever-action design prepared a bullet for firing with one smooth and simple motion, requiring no cartridge biting, no ramrods, and no fumbling for percussion caps. Unlike the three shots a minute that might be hoped for from an exceptionally quick infantryman armed with a Springfield, trials demonstrated that an average shooter could discharge a Henry rifle once every 3 seconds!

An important event occurred on May 10, 1863, on the Jerusalem Turnpike outside of Petersburg, Virginia. By this point in the American Civil War, the overwhelming superiority of a defensive position manned by rifle-armed troops was finally coming to be recognized by most commanders. Southern forces had barricaded the turnpike in order to disrupt a Union cavalry raid deep in Virginian territory. The entrenched troops were ready to turn back any horsemen that might attempt to force the passage. Eventually a small unit of blue-clad troopers did ride up the road, and the Confederates prepared to repel them. Their confidence was boosted by the fact that they appeared to outnumber the Union force by three to one, a rare situation. Despite the disparity, the Yankee cavalry dismounted and, with every fourth man detached as a horse-holder, the petite force made a direct assault on the fortified Rebels. Once firing broke out, an even stranger phenomenon occurred. A storm of lead came pouring into the Confederate breastworks, far beyond what the numbers of advancing men should have been able to produce. Although the Southern forces held a better position, some soldiers became confused and a few of the more superstitious lads reportedly began to suspect demonic forces were at work. The Rebels fell back as the Union troops overran their post.

The devil behind this rare successful frontal attack, of course, was Benjamin Tyler Henry, who had designed the wicked devices carried by the men of the First District of Columbia Cavalry Regiment. The "Dee Cees" were a special unit under the direct command of the provost marshall of the War Department, and because they were an elite force closely connected with the administration, they were provided with the best available equipment including Henry rifles. They were in fact the only regiment in the entire Army of the Potomac who were so well armed, as the $37 average cost for a new gun was considered too exorbitant for regular rank-and-file troops. Nevertheless, individual soldiers and even entire companies dipped into their personal funds to purchase these incredibly rapid-firing guns. Where it was encountered, the Henry earned a fearsome reputation.

According to an oft-repeated legend, one of General William Sherman's troopers who spent several months' worth of his salary to obtain one encountered a group of prisoners who complained that the gun was unfair since it could be loaded on Sunday and shoot all the rest of the week. They were not too far from the truth in this assessment; amongst Northerners, the weapon was nicknamed the "sixteen-shooter." In the South, it was more pointedly known as that "damned Yankee rifle." Impressive as the weapon's reputation was, its actual impact on the war itself was light. By the end of hostilities, the War Department had only officially bought 1,731 of the guns. Still, between that amount and the thousands of private purchases, the Winchester Company's financial success was assured.

A similar repeating weapon that actually saw much more use in combat was the Spencer rifle. Christopher Spencer, a Connecticut gunsmith, patented his basic design in March 1860, the same critical year when both Smith and Wesson's rimfire cartridge and the Henry rifle appeared. Spencer's weapon also used a rimfire cartridge, but in a .56 caliber more closely matching the military's preference for larger bore guns. The Spencer likewise employed a magazine tube, but this feature was set in a different location, behind the lock hidden in the butt of the gunstock. Interestingly enough, the most modern "bullpup" assault rifles likewise hold their magazines in the buttstock. Ammunition was fed into the Spencer's breech from behind. The shorter space of the butt and the larger bore size only allowed seven cartridges to be loaded at a time, as opposed to the Henry's sixteen, yet this was still an enormous improvement over the single-shot breech-loading Sharps—let alone over a muzzle-loading Springfield. Once each weapon was completely empty, the Spencer was actually easier to reload than was the Henry. A soldier with the latter weapon had to unfasten the end of his rifle's magazine tube and individually push in fifteen cartridges. With the Spencer, a latch on the end of the buttstock was unhinged and the empty magazine tube removed. Another preloaded tube was then slid into place and the weapon was ready to fire seven more shots. Solders could carry ten of these tubes in a special leather container called a Blakeslee Quickloader.

This was only part of the reason the military was more supportive of the Spencer than the Henry. The weapon's larger bore size was of course more familiar, and the buttstock enclosed magazine was slightly more durable than the exposed tube of the Henry. The Spencer was a touch more rugged and reliable in general, and some of them were a bit cheaper on the front end. Spencers came in rifle and carbine versions, and the actual average cost of the former over the course of the war was $.50 more per gun than was a Henry. While Spencer charged more for his long rifles, the shorter light carbine version of the same weapon was a steal at $25.40.

Breech-loading carbines were already well established for cavalry use anyway, so Federal forces bought tens of thousands of them. The horsemen were happy to get them. The famous Union cavalry commander John Wilson stated categorically: "There is no doubt that the Spencer Carbine is the best fire-arm yet put into the hands of the soldier, both for economy of ammunition and maximum effect, physical and moral. . . . I have never seen anything else like the confidence inspired by it in the regiments or brigades which have it" (Marcot 1983, 75).

An equally enthusiastic fan was Abraham Lincoln, who met with Spencer and test-fired his weapon in August 1863. The president was so firm in his support for the gun that he personally ordered a reluctant Board of Ordnance to begin purchasing them. Officially they bought 94,196 of the carbines and 12,471 rifles, and thousands more undoubtedly made their way into service through private funds. The rifle performed well at Gettysburg, Chickamauga, Atlanta, Petersburg, and Nashville in particular, where cavalry troops armed with it crushed the Confederate left flank. In addition to earning some of the same nicknames and anecdotes that were applied to the Henry, the Spencer was also reportedly known as the "horizontal shot tower" amongst Confederates.

To a large extent, both the Spencer and Henry rifles represented the beginning of a final major phase in the development of firearms. The self-contained metal-percussion cartridge had allowed these two rifles to emerge, and the same key form of ammunition would propel the gun into its peak level of performance. These repeating rifles had clever designs that continued to reappear in succeeding decades and can even be found in use today, but they had by no means exhausted the potential application of the metallic cartridge. In fact the door was now wide open to a vast range of even more innovative and rapid-firing designs. On one level the firearm was entering its period of maximum potential, but in another sense it was just perhaps beginning to glimpse seniority in the not-too-distant future.

6

Technological Takeoff: The Supremacy of Rapid Fire

◆

THE BREECH-LOADING REVOLVER

One of the more successful examples of the application of the rimfire cartridge to a breechloader design was made by the two men who had played such a critical role in the development of this ammunition, Horace Smith and Daniel Wesson. These two gunsmiths, who had worked hard to perfect the rimfire, were not about to lose out completely to Winchester or Spencer. On the contrary, part of their willingness to hand off the Volcanic design in 1857 came from the fact that they had sensed opportunity elsewhere. Samuel Colt's primary patents on his revolving pistol expired that same year, and Smith and Wesson had big plans in that particular direction. Shortly beforehand they had gone into business with Rollin White, one of Colt's gunsmiths who had unsuccessfully tried to convince his employer to fit his pistols with a breech-loading system. White had taken out a patent on a revolver with the individual chambers bored entirely through the length of the cylinder so that a cartridge could be inserted through the rear.

As it turned out, the basic rimfire design worked well in the Spencer and Henry rifles, but it also served its original purpose as revolver pistol ammunition. Just as those repeating rifles emerged, Smith and Henry began offering light pistols using a very small .22-caliber rimfire cartridge. Their Model #1 employed a unique design in which the front part of the gun could be unhooked and swung upward, allowing the entire cylinder to be removed for reloading. While their .22-caliber pistol was ideal for target shooting, it was too small for military or law enforcement purposes. Smith and Wesson's second model used a slightly larger .32-caliber cartridge, but it too failed to generate much enthusiasm with the army. In the late 1860s they began to experiment with larger cartridges, including the now-famous .44 Henry and similar center-fire variants. By the early 1870s, there was a general movement back toward center-fire cartridges that were now viewed as safer. Larger caliber rimfire cartridges tended to tear themselves apart when firing and were suspected of being more prone to accidentally discharging when jostled while in storage.

In addition to experimenting with new types of ammunition, Smith and Wesson also significantly altered the revolver's design by adopting a variant of Pauly's original hinge-action system. Unsnapping a latch at the rear of the pistol allowed the entire front end of the gun, barrel and cylinder, to drop down and forward. A spring-loaded auto-ejection rod simultaneously pushed all six empty cartridges up and out of the cylinder. This greatly sped up the reloading processes over earlier models and became a hallmark feature of Smith and Wesson revolvers. It was a clear improvement over the Colt system, and one of these models made its way into the hands of Colonel Alexander Gorloff, a Russian military attaché. Gorloff had earlier been impressed by Colt's weapons, but now he became interested in the new design and sent the weapon back home for further testing. It was received so well that in 1871 the tsarist state ordered 20,000 more. The design underwent a few minor revisions, and soon the Smith and Wesson contract with St. Petersburg had swelled to 215,740 revolvers! On a side note, Gorloff's procurement efforts also included new American metallic cartridges and breechloaders designed by Berdan (Bradley 1990, 104–116).

Around this time, White's patent on the rear-loading cylinder expired and other firms became free to experiment with these types of designs. The most successful effort in this endeavor was made by Smith and Wesson's main competitor. In 1871 Colt produced its first breech-loading metallic-cartridge firing pistol, modeled heavily on its earlier cap-and-ball revolvers. This particular gun was not a commercial success, but Colt's next effort would go on to become one of the most famous and distinct revolvers of all times. The Single Action Army Model of 1873 used a solid metal frame around the top of the cylinder that looked similar to the Smith and Wesson but did not unhinge open. Instead, the .45-caliber cartridges were loaded one by one through a hinged feed slot or "gate" in the rear of the cylinder. Empty cartridges were removed in a reverse motion of this action with the assistance of an ejector rod. This was actually a much slower and more laborious process than the one-step Smith and Wesson. In the plus column, however, the gun was rugged, reliable, and inexpensive, retailing for about $15. It was purchased in large numbers all over the country and overseas, but its distinctive curves and rounded pistol grip are today firmly imbedded in the popular imagination of the Wild West. The weapon earned a number of nicknames, including the "equalizer" and the "peacekeeper." Later models were produced well into the next century in a variety of caliber sizes with a number of different barrel lengths.

One of Colt's notable early rivals in Britain was Robert Adams, who produced a line of revolvers in the 1850s that the Royal Army and Royal Navy favored over the American pistol. Their support of the English gunmaker had little to do with nationalism or patriotism but rather was the end result of a patent Adams purchased from a Mr. Frederick Beaumont in 1855. This allowed the Adams revolver to incorporate a feature known as double action. Basically it gave the shooter two options when firing the weapon. If time permitted a shot to be lined up carefully, the soldier pulled back the hammer and cocked the gun with his thumb, much like a Colt or a Smith and Wesson. On the other hand, if speed was paramount, the gun could be fired several times in quick succession by repeatedly pulling on the trigger. This self-cocking mechanism required a stronger tug, which tended to spoil the aim of these rapid shots, but the system was popular enough to give the Adams revolver a certain measure of success against its American competitors. By the 1870s, however, these patents had expired and Colt offered a double-action revolver, as did Smith and Wesson several years later.[1]

Colt would offer one further important innovation that improved the reloading speed in many of its revolvers. In its religious-like avoidance of the Smith and Wesson break-open hinge action, Colt came up with an improvement on its gate-loading mechanism in the new Navy Model 1889 revolver. This pistol had the front of its cylinder fitted to a hinged arm or "crane." After firing, the entire cylinder swung out to the side of the revolver, away from the frame. All six cartridges could then be simultaneously pushed out using an ejector rod mechanism similar to that found in the Smith and Wesson. The latter firm responded in kind in 1896 with a light-caliber revolver that copied Colt's swing-out cylinder. Both systems are still found today in many modern revolvers and have been copied by a great number of manufacturers.

1. As of the date of this publication, the Smith and Wesson Company inexplicably claimed to have invented the double-action mechanism in the 1880s. See "The Smith and Wesson Story" (n.d.).

THE PREDOMINANCE OF THE METALLIC CARTRIDGE

Revolver pistols were not the only firearms to take advantage of the possibilities offered by metallic cartridges in the late nineteenth century. A muzzle-loading tradition that was hundreds of years old was rapidly abandoned by most industrialized nations, who recognized that the future of the firearm lay with breech-loading guns that could utilize the new ammunition. In many cases traditional designs, or even existing firearms themselves, were modified to accept these fancy newfangled rounds. In earlier decades thousands of flintlocks had undergone simple conversions in order to employ percussion caps; it was now time for percussion lock models to endure similar operations.

As a case in point, Britain moved to convert its slow-loading Enfield into something much more lethal, at least as a stopgap measure until a better gun could be designed. By 1864 the British military was well aware of the performance of breech-loading weapons in the American Civil War and publicly advertised for any design that could modify their existing stocks of rifled muskets. The winning invention came appropriately enough from an American, Jacob Snider, who essentially cut a 2.5-inch section off the rear end of the Enfield barrel and refitted a new breechblock to it. This addition had a hinged top that could be flipped up and open to the right, thus exposing the open breech. The rifleman placed a single fully metallic cartridge into the rear of the barrel and snapped the breechblock's top shut again. The modified Enfield kept its external hammer, but this now drove forward a firing pin instead of exploding a percussion cap. After shooting, the soldier was instructed to reopen the breech and turn the firearm slightly upside down while manually pulling the hinged breechblock straight back. This final motion activated an extractor that pushed the spent cartridge out. Since the gun was upside down, gravity did the rest. The design was not an entirely American affair since British Colonel Edward Boxer developed the brass center-fire cartridge that the weapon employed.

This sort of cross-Atlantic international technology transfer was common in the nineteenth century. Ironically, over the course of the next 100 years, the particular percussion-priming system found in the British Boxer cartridge went on to a long career in the United States. A slightly different center-fire primer developed by Berdan, the former U.S. Sharpshooter commander, found greater popularity throughout Europe and the British Commonwealth. The difference between these two priming systems is fairly minute: the former was designed to send a single larger jet

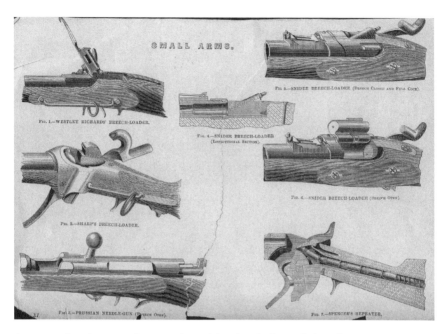

A variety of mid-nineteenth-century breech-loading locks, including the Prussian Needle Gun, Sharps, Snider, and Spencer. © Hulton Archive/Getty Images.

of fire into the cartridge's main powder charge, while the latter created smaller, multiple spouts of flame.

As a nod to the person who added the new breechblock, the modified rifle became known as the Snider-Enfield, and it became the standard service arm for the British Army in a stopgap role while an improved breechloader was being developed. Even after being replaced by superior weapons, the Snider-Enfield continued to find employment with colonial forces in remote parts of the British Empire. In 1867 Russia followed its rival's lead and adopted a similar conversion for its Tula rifled muskets. This modification was made by a Czech inventor, Sylvester Krnka, who came up with a design employing a different hammer and a hinged door that opened to the left rather than the right.

Not to be left behind, the French likewise converted a large number of muzzleloaders in a similar manner. The new breechblocks were fitted to several different models of guns and struck troops using them as reminiscent of a snuffbox, or *Tabatiere* in French. The tabatiere conversion was mechanically closer to the Snider than to Krnka, but experienced a particularly short career. The conversion was only made on largely obsolete firearms that were already being replaced throughout the French Army by the superior bolt-action Chassepot rifle. Compared to the new weapon, the tabatiere guns fired older, heavy bullets that were going out of favor. As good as the Chassepot was, after losing the Franco-Prussian War, an effort was made to improve it even further. The task was accomplished by a certain Captain Basile Gras, who produced a conversion for the Chassepot in 1874 enabling it to use metal center-fire cartridges.

The United States had also come to recognize that the day of paper cartridges was quickly disappearing and sought to convert some of the vast stores of the Springfield rifled muskets they had accumulated in the course of defeating the Confederacy. There was a series of conversions conducted over the next several years, usually with an eye toward economy. For example, the U.S. Ordnance Board saw a similar problem with the large bore that had helped sink the tabatiere conversions. Rather than purchasing new barrels for the Model 1868 Springfield, they initially had the old .58-caliber barrels drilled out and special liners soldered into them in order to reduce the bore size. These conversions and other new models of breechloaders employed a somewhat unique hinged action. Instead of opening to the side, the breech opened upward and forward, thus exposing the bore. This mechanism, which became known as the "trapdoor" amongst troops, remained standard on U.S. infantry rifles into the 1890s. A few were even dragged along to Cuba during the Spanish-American war.

In the meantime, the British army had come up with a replacement for its Snider-Enfield conversions. One of the early inspirations for the new weapon was a breech-loading weapon originally designed in Boston by Henry Peabody, which employed a trigger-guard lever to rotate the breechblock at an angle, rather than straight down as in the Sharps. The breechblock was designed with a shallow channel running along the top that led directly into the rear of the barrel when the trigger guard was pulled downward. This action interested a Swiss gun manufacturer, Friedrich Martini, who removed the external hammer in lieu of an internal firing pin. His system also utilized a separate, longer lever located behind the trigger guard. After looking at 120 different weapons, the British Army decided to adopt the Martini design, but with special rifling made by Alexander Henry of Scotland. Peabody's original contributions were somewhat overlooked, and the weapon became known as the Martini-Henry.

The British wanted the rifle to fire a more modern, smaller, .45-caliber cartridge instead of the larger diameter projectiles of the Snider-Enfield. One problem emerged with this, however. In order to use enough powder to achieve the muzzle velocity that the Ordnance Board wanted, the cartridges would have to be made excessively long, too long in fact to be accommodated by Martini's breechblock. The solution was a modified Boxer cartridge with a bottlelike design. The end section of the cartridge that held the powder charge was of a larger diameter than the forward part that held the bullet. The cartridge could now be made short enough to fit in the Martini action. This short-chamber Boxer-Henry cartridge was one of the earliest examples of the "bottleneck" design that is now widely used in rifle ammunition. This particular cartridge allowed the Martini-Henry to pack a much heavier punch than an earlier .45-caliber gun with a uniformly straight cartridge would have had. Unfortunately the weapon was also known to pack a much heavier kick than most recruits cared to experience, and the gun failed to gain much international support at least partially due to its "excessive recoil" (Butler 1971, 124).

Although trapdoor Springfields and lever-action Martini-Henry rifles used powerful cartridges with long ranges and flat trajectories, their rate of fire left something to be desired. They certainly could be fired more quickly than their muzzle-loading ancestors, but when compared to older Spencer or Henry rifles they were positively sluggish. By rejecting these rapid-firing weapons, in a very real sense the United Kingdom and United States armies had almost taken a backward step in arms procurement. Other interested parties did not follow suit and, in at least one famous case noted below, this would come back to haunt U.S. servicemen in Wyoming.

Despite the success and reliability of their weapons, the Spencer Arms Company went out of business in 1869. This was partially due to an inability to compete with their own secondhand products. Thousands of perfectly functional Spencers had been unloaded by the U.S. government after the end of hostilities, and these retailed for as little as $7.00 each. Because the Henry had been purchased in smaller numbers, there was still a good market for these particular repeating rifles. Winchester also offered new versions of the gun that enjoyed several modifications and improvements. The Model 1866 Winchester was equipped with a wooden fore stock under the barrel that made the hot gun easier to handle after several shots. This was a pretty obvious requirement that the original Henry had failed to employ for some reason. The new repeater also included a loading port on the left side of the gunlock that made topping off the tube magazine with fresh cartridges more convenient. Winchester went on to market other weapons with a variety of slight modifications, but the basic lever-action, tube magazine system remained largely unchanged.

Among Winchester's more intriguing designs were the poorly named Winchester Muskets. These versions contained several nods to military tradition such as a longer fore stock that was cosmetically similar to an infantry long-arm, and a barrel designed to accept a bayonet. The guns were still fast repeaters, and it is hard to imagine any enemy soldiers managing to get close enough to one to actually be bayoneted by it. Still, the cash-strapped U.S. military choose to rely upon the thrifty Springfield, much to the misfortune of the five companies of the U.S. Seventh Cavalry that were wiped out at the Battle of Little Bighorn. It appears that a significant number of the Lakota-led alliance they encountered were armed with faster-firing Henry and Winchester rifles, creating an unusual reverse situation. Unlike most encounters of the era, in this particular fight the indigenous resistance may have been better armed than the colonizing power.

Although the U.S. military was not willing to purchase the lever-action Winchesters, their presence at Little Bighorn shows that plenty of others were. Tsarist Russia was not the only state to show an interest in new American firearms as the rival Turkish Ottoman Empire placed orders for some 50,000 Winchesters by 1871. These were put to use against the former power in 1877 at the siege of Plevna, where an outnumbered Turkish force held off a major Russian advance for six months. The Turkish army gave a significant portion of the credit for this success to their repeater rifles and quickly ordered 140,000 more.

The guns were also far more popular with the civilian market than Oliver Winchester's shirts had ever been, and when he died in 1881, his widow Sarah reportedly gained some $20 million and nearly half of the stock in the Winchester Repeating Arms Company. Much of this wealth was poured into the notorious "Mystery House" that she began building in 1884. Sarah apparently believed that she was being haunted by the spirits of people killed by her husband's weapons and that the only way to deal with the ghosts of Confederates, cowboys, and Cossacks was to leave New Haven, move west to California, and build a massive mansion. Whether it was constructed to house the phantoms or to just trap and confuse them is a bit unclear, but it does seem that Sarah did not want its construction to be completed in her lifetime. The end result was a 160-room monstrosity that covered 6 acres of land and was filled with architectural oddities, including doors and staircases that led nowhere.

THE MODERN BOLT-ACTION RIFLE

Unlike the halls of the Mystery House, the great powers of Europe had fairly clear ideas about the directions in which they wished their soldiers' firearms to move. In particular the Winchester's tube magazine inspired a similar system in several other firearms. In 1868 the Swiss army adopted a repeating rifle designed by Friedrich Vetterli that incorporated this mechanism along with the cartridge "elevator" used in the Henry rifle. While these elements were similar to those of the Winchester, they replaced the lever with a bolt action like that found on the von Dreyse needle gun. This design was later copied by Austria-Hungary, Germany, France, and even Japan.

Although all of these bolt-action, tube-loading rifles were fairly similar, one of the most famous was designed by Peter Paul Mauser. In 1871 the company he and his brother Wilhelm founded won a competition to replace the von Dreyse with a design using metallic cartridges. Their rifle was a basic bolt-action, 11mm (about .43-inch), single-shot breechloader. While it was a much more reliable weapon, its actual rate of fire remained little changed from that of the original 1838 needle gun. The "Plevna Delay" had made it clear that repeating rifles were the way of the future, and Mauser eventually improved his design in 1884. He copied Vetterli's basic system and added a tube magazine to the fore stock that fed cartridges into the breech when the bolt action was operated. Although the gun may not have been the most innovative weapon in the world, it did earn a reputation for unmatched durability that earlier repeaters did not enjoy. It is estimated that as many as 1 million of Mauser's 1884 version were eventually built, finding great popularity in the international market.

France also appreciated this design, and in 1878 they began adopting limited numbers of a bolt-action, tube-magazine rifle developed by Alfred von Kropatschek for special service units. Eventually, however, in 1886 the army settled on a similar rifle that was tested under the supervision of Colonel Nicholas Lebel. Although he did not feel he deserved any credit for the actual development of the weapon, it came to be known after him. The basic bolt-action design of the Lebel was nothing particularly new, but its ammunition certainly was. First of all, the caliber of the gun was significantly smaller than the generally accepted 11mm bullet. The Lebel fired a projectile that was only 8mm (about .31 inch) in diameter, yet still was considered sufficiently powerful.

The secret behind its punch was the result of a propellant developed by Paul Vielle in 1884. For centuries, firearms had used the same basic gunpowder to push bullets down their barrels. The exact percentage of ingredients could vary widely, but the three basic ingredients of saltpeter, sulphur, and charcoal used in the Heilungchiang gun of 1288 were still being ignited in the modern repeaters nearly 600 years later. However, experiments with fibrous material like cotton treated with nitric acid were producing interesting results. Scientists like Vielle and Alfred Noble would eventually develop stable explosives from these products, which were commonly known as "guncotton" or "nitrocellulose." Vielle managed to come up with a new propellant for firearms.

Vielle's particular formula was based on gelatinized nitrocellulose mixed with ether and alcohol. The mixture was rolled into flat sheets, dried, and cut into small flakes or granules. These could be loaded into a cartridge just like gunpowder and similarly ignited by fulminate. Vielle called his new substance "Powder B," but it and other similar propellants came to be simply known as "smokeless" powder. In fact these formulas *do* produce smoke, but significantly less than the older gunpowder. This is an obvious advantage because a hidden shooter was less likely to reveal his position. Thousands of guns firing the older powders in a battle could also produce a haze that impaired general visibility, although this effect admittedly could be mitigated by even a light breeze. On the downside, the fancy new powder also took some of the visual fun out of shooting a gun. It should also be noted that because early smokeless powders were of a light color, people began to refer to the older charcoal-based gunpowder as "black powder," a designation it still carries today.

Beyond its low-smoke properties, Powder B was also a more efficient propellant. While good black powder is made by carefully and thoroughly mixing its primary ingredients, a smokeless powder's reactant materials are actually combined on a molecular level. The ignition of this matter is thus much faster, as the reaction can move more quickly through the chemical. It also burns more thoroughly, which is why it produced significantly less smoke and, incidentally, left less powder residue in the barrel. If nitrocellulose is manufactured with a very high ratio of nitrates to a lower amount of cotton or wood pulp, the reaction is so thorough and violent that the shock wave traveling through the compound travels faster than the speed of sound. As indicated in Chapter 1, this is a characteristic that modern science defines as a "detonation." Noble's particular invention, dynamite, for example, acted in this fashion. Vielle's treated nitrocellulose, on the other hand, burned a little slower and technically speaking deflagrated, but it still packed a more powerful punch than the less efficient black powder.

What all this meant for the Lebel rifle was that its bullet could be blown down the barrel at a much faster rate than the projectiles of other contemporary firearms. The force that a projectile has on an object when it impacts with it is a combination of both mass and speed. A relatively slow-moving thing with a large mass, such as a 20-ton bolder rolling down a hill, clearly has the potential to do a lot of harm. Likewise, a tiny object traveling at extremely high speed is similarly lethal. For example, space debris from old satellites and rocket ships orbit the earth at an average of 17,500 miles an hour. An object as small as a common bolt traveling at such speed could cause serious damage if it were to hit a spacecraft. As problematic as this particular law of physics is for twenty-first-century space programs, it created opportunities for nineteenth-century firearm designers. A small bullet would be just as capable of inflicting a similar level of injury as a large one, as long as it was traveling at a relatively higher velocity. There would be some differences in the specific physical nature of the wounds inflicted, but for all practical purposes the effects would be comparable. It was the stronger Powder B that thus allowed the Lebel to employ the smaller caliber bullets. Contemporary black powder weapons typically propelled larger ammunition of 11mm (.43 inch) or .45 inch (11.43mm) with a muzzle velocity of 1,000–1,200 feet per second. In contrast, the smaller 8mm Lebel bullet hurled out of a gun barrel at over 2,000 feet per second.

Impressive as this velocity was, it created another problem. Small projectiles traveling at this speed tend to tumble in flight, utterly spoiling their accuracy. It was found, however, that the bullet could be stabilized by increasing its rate of spin. The way to achieve this effect was to make the Lebel's rifling "tighter." The rifle grooves inside the barrel made a complete turn or rotation within a 9.4-inch length of the barrel. This was shorter than the rifling used in earlier guns, but prompted yet *another* challenge to be hurdled.

A soft lead bullet traveling at an extremely high velocity across such tight rifling tends to tear itself up on the edges of the grooves. Such a distorted projectile likewise loses accuracy. Since the bullet was being shredded along the inside of the barrel, it would also leave relatively more lead fouling inside. A Swiss captain named Edouard Rubin had been experimenting with different metals in ammunition and came up with a solution. Lead was still necessary to add weight to the projectile, but if a thin metal jacket was wrapped around the bullet's exterior, the problems of a damaged projectile and lead fouling were largely eliminated. The best metal for this job had to be pliable enough to grasp the gun's rifling, but sufficiently resilient to avoid being torn apart. A mixture of copper and nickel called "cupronickel" seemed to do the trick. A bullet jacketed with this material was incorporated into the Lebel cartridge's design and became another standard element of future firearms.

The Lebel rifle was just one in a series of important breakthroughs in a rapid age of firearm developments. In many ways, it is probably better to think of the Lebel cartridge as the real breakthrough and the gun as an accessory to it. The design was so successful that some 4 million of these particular firearms were built between 1886 and 1917. The rifle saw service in numerous actions around the world and remained the standard French service arm well into World War I.

While the Lebel design represented a superior infantry firearm, rival states developed additional modifications to the bolt-action infantry rifle. Most of these revolved around improving the loading mechanism. One drawback to tube-magazine rifles was the manner in which they were replenished with fresh ammunition. Once the eight cartridges in a Lebel's tube magazine had been fired off, the gun had to be reloaded cartridge by cartridge. Some feared that this could potentially be a dangerous drawback if an enemy made an aggressive attack. After eight enemy soldiers had been dispatched, a ninth one just might be able to shoot the helpless soldier while he was fumbling with a loose cartridge. Such concerns were probably exaggerated as the Lebel rifle managed to kill thousands of Germans during World War I, even with its "slow" reloading system. Firearm designs were already beginning to reach something of an overkill approach as it was. In any event, a "solution" to this imagined problem was already underway.

The same year the Lebel was adopted, the Austro-Hungarian army had similarly selected a weapon designed by Ferdinand von Mannlicher. Its ammunition was clearly inferior to that of the Lebel, and the Austrians almost immediately began to redesign it in the wake of the revolutionary French gun. Mannlicher's feed mechanism, however, attracted a great deal of international attention. Rather than using a tube, the Mannlicher contained a hollow magazine directly under the breechblock-bolt mechanism. The position of the magazine under the bolt limited the number of cartridges that could fit in this space to a mere five. These five were attached, one on top of the other, to a metal loader popularly known today as a "clip." To load the Mannlicher rifle itself, the bolt was pulled back and the loader was dropped into the empty magazine space. As the bolt action was worked, it drew fresh cartridges that were pushed up into position from the clip by a simple spring. Once all five cartridges had been used, the empty loader dropped out through an aperture in the bottom of the magazine. The gun had to be recharged after five shots, as opposed to the Lebel's eight, but the process was much faster. An individual soldier could easily carry a dozen or more of these preloaded clips, especially when they were made petite and light for a smaller Lebel-like cartridge.

A number of countries came up with versions of this clip-loading or "packet-loading" system, as it was variously known. Perhaps the most famous and highly perfected designs emerged out of Germany. The Model 1888 rifle basically took the older Mauser design and refitted it with the Mannlicher clip system. It employed a cartridge similar to the Lebel in the then-unusual caliber of 7.92mm. This ammunition also dispensed with the small "rim" around the base of the cartridge that had been typical of previous metallic designs. The little modification reduced the chance of cartridges catching on the edge of each other or gun components and causing feeding jams. A year later, Mauser modified this system even further for use by the Belgian army. He introduced a version of the clip known as a "charger" or sometimes as a "strip clip." This thin metallic device held five cartridges by the rear, one on top of the other, similar to a Mannlicher clip. When loading, however, the strip clip itself did not go into the gun itself but was fitted in place above the open magazine. The shooter could then use his or her thumb to slide the cartridges down off the charger into the weapon. The empty charger was then removed, and the gun was ready to fire in the traditional manner. The supposed major advantage to this system was that if a soldier fired one or two shots, and then encountered a break in the action, the magazine could be "topped off" by pushing in a couple of extra cartridges. Useful as this function might have been, it does seem like much ado over little. Nevertheless, not only were the Belgians impressed by the system but so were the Americans, Germans, Japanese, Swedes, and people of other nations who adopted versions of it. A large element of warfare is psychological, and if a soldier felt more comfortable by being able to top off his weapon, than perhaps the Mauser design was a worthwhile innovation.

There were yet other variations on the basic bolt-action approach. For example, the magazine did not have to be set directly beneath the bolt, but could be arranged in other positions. An odd approach like this was made by the team of Colonel Ole Hermann Johannes Krag and Mr. Erik Jorgenson, who designed a bolt-action rifle for the Danish army. The opening to the magazine on this weapon jutted slightly out to the right side of the bolt action and was fitted with a trap door on the end. Although a round could be placed directly through the breech into the barrel, the Krag-Jorgenson was generally loaded by means of the trapdoor. The cartridges then lined up one next to the other underneath the bolt. As the gun was repeatedly fired, a spring fixed to the magazine's trapdoor pushed them one by one under the bolt and fed them into the breech from underneath and to the *left* of the breechblock. The big advantage that the Krag-Jorgenson offered (and that impressed U.S. officials) was that the magazine could be topped off through its door without even opening the breech (Hogg 1996, 74). On the downside, Krag-Jorgenson cartridges were loaded one by one and did not employ a clip or a charger. The U.S. military used the weapon briefly in the Spanish-American War, but shortly thereafter they decided to adopt the Mauser system in their new .30-caliber (7.62mm) Model 1903 Springfield rifle.

Britain likewise decided to abandon the single-shot Martini-Henry in favor of a bolt-action weapon, ultimately adopting a version in 1888 that was slightly different from the standard Mauser action. This was an amalgamated weapon that included a barrel designed by William Metford and a bolt-action system designed by James Lee. In the long run, Metford's contribution was proven to be less significant than Lee's magazine design. The Lee-Metford employed a clever box-shaped magazine that was situated in approximately the same place as that of the Mauser, but was *detachable*. This eight-shot magazine could be removed and replaced through a slot in the bottom of the weapon under the bolt action. These removable box magazines also became informally (and, some would argue, inappropriately) known as "clips" because they were used in a similar manner and performed a similar function as the Mannlicher and Mauser devices. A British soldier could carry a number of these preloaded box magazines, just as the German soldier carried preloaded clips. Unlike the Mauser or Springfield, the gun fired a more powerful .303-caliber (7.69mm) cartridge. As it turned out, the Metford barrel had trouble holding up to the wear and tear of the .303 and had to be replaced by a more durable tube made at the Enfield armory. The rifle was renamed the Lee-Enfield, and later models became simply known as

A Women's Royal Naval Air Service officer examines a Lee-Enfield rifle. © Hulton Archive/Getty Images.

"Enfields." Although the gun used a late-nineteenth-century bolt action, its detachable box magazine would become a standard feature in faster firing firearms developed throughout the twentieth century.

THE MANUALLY OPERATED MACHINE GUN

As rapid as these bolt- and lever-action reloading systems were, the firearm had one last major series of mutations to undergo before reaching its current state of equilibrium. The desire to increase rates of fire was clearly a major imperative behind many of the advances in firearm technology over the course of the nineteenth century. Just as the century was not yet over, however, neither were all the possibilities for rapid-firing gun designs exhausted. This age of industrialization, machines, and automation would implant itself more directly upon the story of firearms than the simple means by which these weapons were produced. Now efforts were undertaken that were intended to turn the gun itself into a kind of machine. The end results were weapons known appropriately enough as *machine guns*, devices capable of sustained automatic fire.

Like so many nineteenth-century firearm developments, the machine gun had much older ancestors that were intended to accomplish similar tasks or to work upon similar principles. The late fifteenth century saw the advent of "organ guns," so called because of their visual similarity to church organs. These devices consisted of multiple barrels fixed to a single stock, or more often to a wheeled carriage. The barrels were typically arranged so as to point in a single direction but sometimes were splayed outward in the duck's foot pattern. Da Vinci made sketches of different designs using both patterns. Organ guns typically fired ammunition sized anywhere from .75 to 2 inches in diameter and might carry ten, twenty, thirty, or even more gun barrels. These weapons were made so that a single ignition traveled through passages cut in the rears of the barrels, firing each in succession. Some complicated nineteenth-century organ guns also employed the "roman candle" technique noted in Chapter 5, allowing a single weapon to deliver hundreds of shots (Reid 1984, 66). These devices must have been as cumbersome to employ and as slow to reload as they were spectacular to watch when finally fired.

Another interesting precursor to the machine gun was James Puckle's "Defence." This early-eighteenth-century experimental English flintlock weapon was designed to operate upon the revolver principle. Because it used a single barrel, it was less weighty than an organ gun and instead was mounted on a light tripod. The gun used a nine-shot cylinder that was rotated into place by a rear-facing handle. After each shot, a new loaded chamber could be cranked into position and the weapon discharged again. Its overall rate of fire was lower than an organ gun, but since extra cylinders could be kept on hand, the reloading process was nowhere near as laborious. In one trial, the gun discharged sixty-three shots in 7 minutes. In a crusade-like sentiment, another version of the gun was reportedly designed to be loaded with special cylinders that could fire square bullets for use against Turks and (presumably) other non-Christians. For some reason, the device failed to arouse much interest with army officials.

Puckle's design is also of interest due to its superficial similarity to some of the first successful hand-operated machine guns created over 100 years later. The very term "machine gun" is clearly linked to the nineteenth-century fascination with new technological innovations and industrial production techniques. Just as it triggered a variety of rifle and carbine designs, the American Civil War also encouraged inventors to try and develop new automated guns as well. One of the more promising designs, developed by Wilson Agar, used a rear crank handle to drive the firing mechanism. Unlike Puckle's Defence, it employed a funnel-shaped magazine or "hopper" instead of a revolver cylinder. Prior to action, a number of small reusable metallic tubes were individually preloaded, each with a .58-caliber paper cartridge. A separate percussion cap was fitted to a nipple fixed in the rear of each of these tubes, which were in turn loaded into the hopper. Cranking the rear handle then fed the cartridges into the breech and fired them one by one. The mechanism's similarity to a more common household implement led to its informal nickname, the "coffee mill." The Agar was capable of firing in excess of 120 shots per minute but it was prone to overheating, a factor that caused some concern.

A different manually operated machine gun created by Dr. Richard Gatling managed to avoid this particular problem. Like Maynard, Gatling was a dentist with more interest in guns than gums, and in 1861 he patented another hand-cranked, revolving, rapid-fire weapon. Although he never acknowledged it, the inventor almost certainly borrowed some inspiration and elements of his design from the Agar and from a lesser known design by Ezra Ripley of Troy, New York (Wahl and Toppel 1965, 9–11). For example, Gatling's initial models used reusable loading tubes much like those employed in the Agar, although he later replaced this somewhat complicated system after 1865 with regular metallic cartridges.

The Gatling employed the revolver mechanism seen on Puckle's Defence, but solved the overheating problem by using more than one barrel. Visually, the weapon looked a bit like a giant pepperbox pistol mated to an artillery piece. Like the pepperbox, there were no separate "cylinders" behind a single tube but rather a grouping of half a dozen barrels each with its own breech mechanism. The whole structure was obviously quite heavy and needed to be supported by a wheeled carriage, similar to that of a light cannon.

The cumbersome size of the device was counterbalanced by its performance. Since each barrel was only receiving one-sixth of the heat of a single barrel, relatively speaking, the rate of fire could be much higher than that of an Agar. The first Gatling guns used a loose ammunition hopper, but this was later replaced with a detachable magazine system set above the breech. These were of various shapes, including a round "drum" design and a flat curved "banana clip" magazine anticipating those found on later assault rifles. As each barrel revolved past the magazine, it picked up a cartridge that was then fired at the bottom of the rotation cycle. By the time the barrel had revolved to pick up a fresh cartridge, the original empty case had been expelled. The advantage of heat dissipation that this multiple-barrel system offered permitted the gun to fire as fast as the handle could be cranked around and empty magazines

A late-nineteenth-century Gatling gun. © Hulton Archive/Getty Images.

could be replaced. One version, the Model 1883, was capable of firing as many as *1,500* rounds per minute.

Gatling's design is widely credited as being the first really successful machine gun. The weapon proved to be highly dependable and was adopted for military service by a number of countries. Gatling guns were produced in a variety of models sporting anything from five to ten barrels and in caliber sizes up to 1 inch in diameter. Some later versions were compact enough to be mounted on a tripod, a naval gun swivel, or even the back of a camel. Most Gatling guns, however, remained fairly sizable pieces of machinery that had to be lugged around like artillery. In a notorious example of bad judgment, Lieutenant Colonel George Custer reportedly decided not to bring four Gatling guns with him prior to the Battle of Little Bighorn for fear that the cumbersome wheeled weapons would slow down his command. After some similar initial setbacks, Lord Chelmsford made sure to bring several of the guns with him at the more successful Battle of Ulundi during the Zulu War three years later. Many Gatling guns remained in general usage through the 1890s before being replaced by superior automatic weapons.

Incidentally, Gatling's technique of dispersing heat remains one of the most effective ever devised. In 1893 an experimental Gatling gun run by electricity discharged an astounding 3,000 rounds per minute. This volume of firepower was on a scale that the U.S. Air Force was interested in obtaining in the earliest part of the Cold War. General Electric was given the contract for this so-called Project Vulcan and eventually developed a line of Vulcan guns and "Mini-guns" largely based on the Gatling system. Some of these modern incarnations are capable of firing a mind-boggling *7,000* shots per minute, which translates to in excess of 100 shots per second!

As successful as Gatling's design was, it appeared during an explosive period of innovation for the firearm, as others actively tried to develop manual machine guns. Belgian Captain Fafschamps and engineer Josef Montigny in the 1850s and 1860s developed a similar device. Their system so impressed the French army that they adopted it and called it the Mitrailleuse after *mitraille*, a French term for small projectile ammunition fired from a cannon (Hicks 1941, 104–105). Montigny's Mitrailleuse was also a hand-cranked gun that used multiple barrels and was mounted on an artillery carriage, but it differed from the Gatling gun in several respects. First of all, it carried a total of thirty-seven small barrels, none of which rotated. The gun was readied for firing with a preloaded flat metallic magazine holding thirty-seven cartridges arranged to fit in the corresponding barrels' breeches. Because it too used more than one barrel, the weapon was not easily subject to overheating and had a rate of fire that actually exceeded that of the first Gatlings. The gun did not enjoy quite as much international popularity as did the other weapon, perhaps due to an unfair association with France's defeat in the Franco-Prussian War of 1870–1871.

Another manually operated machine gun employing multiple barrels was designed in Sweden by Helge Palmcrantz and was financed by industrialist submarine experimenter Torsten Nordenfelt. Money apparently spoke louder than actions in this particular relationship, and when the weapon was introduced in the early 1880s it was called the "Nordenfelt." The machine gun differed most noticeably from a Gatling gun in that its barrels were lined up horizontally, side by side. Like that earlier machine gun, the number of actual barrels in a Nordenfelt varied from model to model. Guns were developed using two, three, four, five, and even ten barrels in caliber sizes ranging anywhere from .45 up to 2 inches. The latter projectiles were used in two-barreled versions that are better classified as light artillery than as machine guns. Like the Gatling, the Nordenfelt used a top-loading magazine, but this was divided into subsections that corresponded with each of the barrels and fed cartridges into the appropriate breech mechanism. The gun was actually discharged by pushing a side lever back and forth, and through this simple action some of the models were capable of exceptionally high rates of fire.

As advanced as these manually operated machine guns were, they would be pushed aside in a few short decades by more automated versions. The new guns would initially look a little like the older models, being large carriage-mounted devices. They differed in a key mode of operation, however. Rather than relying upon cranks or handles to drive the firing system, the next generation of machine guns would be fully automatic. Simply pressing a trigger would cause the gun to spit out a rapid, continuous spray of bullets. This innovation would usher in the firearm's last major period of development.

THE AUTOMATIC MACHINE GUN

The first automatic machine guns that were developed shortly before the beginning of the twentieth century could fire out hundreds of bullets a minute by simply pressing a trigger. They accomplished this task through two somewhat different principles. In each method, a by-product of the cartridge's explosion drove the mechanism of the gun. The two systems would have a profound impact, not only in the development of future heavy machine guns but also upon most common automatic and semiautomatic firearms in use today.

Perhaps the most lasting and revolutionary firearm of the late-nineteenth century was a machine gun developed by Hiram Maxim. Maxim was another

American-born inventor who, like Gatling, applied his energies to a variety of projects but had most success with firearms. He set up a workshop in London in the early 1880s and began trying to develop an automatic machine gun after reportedly being advised that the best way to get rich was to invent something to allow Europeans to kill each other more easily. The thing that set Maxim's work apart was that he decided to use the energy of the exploding charge itself to drive the mechanism of his weapon, rather than relying upon hand operation. When the propellant in a charge is ignited, the explosion does not travel exclusively down the barrel, but initially pushes outward in all different directions. The metal barrel and breechblock redirects much of this power to drive the bullet toward the muzzle. Some of this energy travels in the opposite direction, however, shoving the gun backwards into the shoulder of the firer. This recoil is an old but highly unwelcome feature of most firearms. Certain exceptionally notorious nineteenth-century Belgian weapons subject to this phenomenon were nicknamed "mule-kickers" for obvious reasons. Maxim sought to employ this mulish element of the firearm to drive his particular machine.

After a few years of experimentation and building prototypes, Maxim had a gun ready to present to the public in 1885. The action of this gun was driven by each shot's recoil, which pushed both the barrel and breechblock

Hiram Maxim with one of his famous automatic machine guns. This particular model has a detachable wheeled carriage. © Hulton Archive/Getty Images.

slightly backward, opening the latter enough to allow an empty cartridge to fling itself out of the gun. Fresh ammunition was fed into the side of the breechblock not by a metallic magazine, but rather by means of an ammunition belt. The Maxim gun used a cloth version with small pockets into which the cartridges were fixed. The force of the Maxim's initial shot also served to operate another mechanism that pulled the belt into the breechblock and lined up a fresh round. After the first empty cartridge was flung clear, a recoil spring helped drive the second forward from the belt into the breech. The first shot had to be cocked by hand, but after that the process was fully automatic, and hundreds of rounds could be fired by keeping the trigger held down. The shooter could thus concentrate more fully upon aiming rather than cranking a handle.

Unlike the Gatling or Nordenfelt, this machine gun employed but a single barrel. The problem of overheating was clearly a potentially acute one, and it was addressed by an ingenious water jacket. At one point in 1865 Gatling had explored the idea of cooling an experimental four-barrel version of his machine gun by fixing metal water-filled containers around each barrel (Edwards 1962, 239). He later abandoned this idea in favor of adding more barrels to disperse the damage the heat could do to any particular one, but the basic concept remained a sound one. Maxim used the same technique to cool his gun. The water jacket was simply a large metal tube that surrounded the weapon's barrel and was filled with cold water. One potential problem with this was that as the water heated up, steam pressure inside the jacket increased. This vapor was released through a valve, but it had the potential to give one's position away to an enemy. The earliest Maxim guns still used black powder, so this was comparatively less of a concern, but later versions of water-cooled machine guns vented their steam into rubber tubing that was attached to a separate reclamation reservoir. Once the steam had condensed back into water, this could periodically be fed back into the jacket. The cooling system was pretty bulky, but was still more manageable than the multiple heavy barrels of a Gatling gun. It was also not visually streamlined (pardon the pun), but it served its purpose well enough and allowed the Maxim to fire at a brisk rate of about 500–600 rounds per minute with relatively little fear of overheating.

The Maxim quickly proved its potential both in military trials and in colonial service. Supported by new bolt-action rifles, the Maxim clearly showed that the relative balance in worldwide firearms technology that had characterized the matchlock and flintlock eras was over, at least for the time being. In 1893 in southern Africa, a British colonial police unit of about fifty men armed with four Maxims and two other machine guns fought off a force of 5,000 Ndebele (or Matabele) fighters. In less than two hours,

3,000 members of the resistance were dead. This lack of technological parity, combined with advances in the prevention of malaria, allowed Europeans to quickly conquer Africa and much of Asia in a mad scramble conducted over the last two decades of the century (Headrick 1981, 115–124). Further dramatic evidence of this weapon's power came in 1898 at Omdurman in the Sudan, where the forces of a charismatic Islamic leader known as the Khalifa attacked the Anglo-Egyptian army of Lord Kitchener. The Sudanese suffered somewhere between 25,000 and 30,000 casualties in a matter of hours. Many of the losses were attributed to the half-dozen or so Maxim guns supporting the British defenses. Such a lopsided battle caused the French-born, London-based poet Hilaire Belloc express gratitude for the fact that "Whatever happens, we have got / the Maxim gun and they have not." Belloc's lines are quite appropriate and clever when the "they" was applied to poorly armed Africans or Asians. What seemed to have been forgotten was that *other Europeans* had the Maxim gun too and would rack up appalling casualties by using it on each other during World War I (Ellis 1986, 111–130). Thanks to Maxim and his ilk, a century after the pageantry of the Napoleonic era, warfare had been reduced to a haze of filthy gray-and-brown-clad millions huddling in muddy trenches trying to avoid the mass-industrialized death that ruled no-man's-land.

As popular as recoil-operated Maxims would continue to be, a second machine gun using a different system of automation had made an appearance by the close of the century. This was the brainchild of yet another American inventor, John Moses Browning. Even though Browning had no training in dental services, as the son of a Utah gunsmith he proved to be one of the most prolific and innovative firearms designers in history. In particular he developed a number of single-shot breechloaders, lever-action repeaters guns, and "slide" or "pump-action" firearms for Winchester that could be reloaded by pushing the front stock forward and backward. As clever as these initial designs were, they generally involved minor variations on previously developed innovations.

However, in one circumstance, Browning came up with something entirely new and innovative. According to Browning, one day he noticed how the muzzle blast from one of his guns pushed away nearby brush and weeds as it was fired. This purportedly gave him the idea that there was extra, "unused" energy being released from the muzzle that might be incorporated to drive a gun's firing mechanism. Actually a number of inventors had tried to harness this force since midcentury, but Browning's story has since passed into legend. It is also possible that he was simply trying to come up with a way to duplicate the performance of Maxim's guns while avoiding his patent attorneys. Instead of relying upon the simple recoil power of the shot, Browning's intention was to tap into the expansion power of the gas created when the cartridge's propellant ignited. He began to build experimental firearms that could make use of this force to rechamber fresh rounds after each shot. The end product of these experiments was a design for a fully automatic machine gun that failed to interest Winchester. Instead, it was snatched up by the competition and emerged as the Colt Model 1895 Automatic Machine Gun.

This machine gun had a small hole drilled in the bottom of the barrel toward the muzzle. Some of the expanding gases driving the bullet were diverted into this opening, where they pushed down a lever that in turn opened the breech. A spring forced the arm back into position and rechambered a new round. Fresh ammunition was fed into the mechanism basically through the same belt system used by the Maxim. The up-and-down motion of the lever arm also gave rise to a curious nickname. Under field conditions, on occasion the swinging arm hit the ground and threw up dirt. The nickname "potato digger" thus came into use to describe Browning's first machine gun.

Beyond looking odd during firing, the long, laborious action of the gun also provided a relative low rate of fire, only about 400 rounds a minute. In practice, though, this was not an enormous drawback. Part of the secret behind the potato digger's light weight was the fact that it carried only a single barrel and no water jacket. Browning made the barrel itself extra thick in an effort to help disperse heat, but it certainly was preferable for a gun like this to have a lower rate of fire. As a further precaution, soldiers were trained to shoot in shorter, rather than longer, bursts of fire. The weapons proved to be reasonably reliable under adverse conditions, and the gas operation provided for a smooth firing action. They were also considerably lighter than other machine guns and could be mounted on a simple portable tripod if desired. Colt produced several different versions to match common ammunition types, such as that used in the Krag-Jorgenson, the 1903 Springfield, and some of the Mauser rifles.

In their work on machine guns, Maxim and Browning each had uncovered basic principles that would drive firearm development for the next century. However, one writer, David Armstrong, noted that Browning's patent on a gas-operated weapon in 1890 marked the end of the basic development of the machine gun. "Subsequent modifications in the design of the weapon reflected advances in metallurgy as well as new ideas about how it should be employed" (Armstrong 1982, 78). This point is a very good one, and while Armstrong was specifically reviewing machine guns, his statement could likewise be applied to firearms as a whole. Alternations and innovations in weight, size, design, bullet velocity, and a host of other features would continue to be made, but most new successful firearms would employ some variation or combination of recoil and/or gas operation. In a sense, the firearm had reached an apogee of development by the dawn of the twentieth century. Nevertheless, hundreds of new automatic and semiautomatic pistols, machine guns, submachine guns, and rifles would still be produced that used one, the other, or both of these systems. Once Maxim and Browning had demonstrated the potential automotive power of the cartridge, numerous gun designers would come to incorporate it into new weaponry.

7

Apogee and Equilibrium: Automation

◆

THE SEMIAUTOMATIC PISTOL

One early object of experimentation was the pistol, and the end result was a new class of handgun that could keep firing without being cocked after each shot. Because the trigger has to be pulled for every round, the system is not quite fully automatic like a machine gun, but rather is considered semiautomatic. A double-action revolver works in a similar manner, but somehow retained its older classification. It is not entirely clear who invented the very first semiautomatic pistol, but an important early claimant is the Austrian Joseph Laumann, who in 1892 patented a weapon called the "Schonberger." Like other later pistols and submachine guns, this weapon used the so-called blowback system, which takes advantage of both recoil and gas operation; thus it is something of a hybrid combination of the two forms (Chinn 1956, 3). In this modified version, the sliding bolt does not actually lock in place like it does in a bolt-action weapon or in the Maxim gun. The heavier weight of the bolt initially holds it in place as the bullet starts to travel. Eventually pressure in the barrel also drives back the bolt and flings the empty cartridge case away, but only after the bullet has left the gun and pressure in the barrel has begun to drop. Generally in the blowback system only the bolt itself moves backwards, not the bolt and barrel together as in recoil operation. While common sense suggests that this should not be a particularly important

Technician examining pistols with attachable butt stocks. He is holding a long-barreled semiautomatic Luger and has a similar Mauser lying near his left hand. © Hulton Archive/Getty Images.

distinction, to gunsmiths and technology aficionados it is a big deal worthy of much dither and blather. One important element of the blowback system to note is that it tends to be found in light-caliber weapons such as pistols rather than in high-velocity rifles. The cartridges in these longarms are simply too powerful to safely employ an unlocked breech mechanism. A single bolt heavy enough to sufficiently retard the blowback of a .30-caliber (7.62mm) rifle cartridge would have to weigh about three and half times the total weight of a typical weapon in this class! While the Schonberger pioneered the blowback system, the gun did not prove to be the most financially successful venture for its producer, the Steyr firm of Austria. It also was one of the most visually unappealing firearm designs ever created, concealing its magazine in an odd rounded protrusion in front of the trigger that looked more like an mutated biological protrusion than a precision machine component.

A different semiautomatic pistol designed by Hugo Borchardt enjoyed a little more luck. Borchardt had worked for Winchester in the United States for a number of years before returning to his native Germany and taking up employment with the Ludwig Leowe firm of Berlin. He produced an innovative gun in 1893 that contained some of the same key elements found in modern semiautomatic pistols. In particular, the weapon had a clever detachable magazine system similar to that found in the Lee-Enfield, but hidden in the pistol grip. The 7.63mm (.30-inch) cartridges were stacked up with one on top of the other and pushed upwards into the breechblock by a spring, much like other magazine-fed rifles. In this case, however, the shooter was relieved of the duty of rechambering each shot as required in a bolt-action gunlock. This job was instead performed by a heavy spring mechanism set in a semicircular housing in the rear of the pistol behind the magazine. This was fixed to the bolt by a hinged arm or "toggle" designed to bend in an upward motion. The recoil from the first shot drove back the breech block, recocked the firing pin, and threw out the empty cartridge, all before the rear spring reasserted itself and pushed the bolt back forward. As it reclosed, it picked up a fresh cartridge from the magazine and drove it into the firing chamber. The system was similar to that used in the Maxim, except that in this case the shooter had to repull the trigger after each shot. Once eight rounds were fired, the empty magazine could be replaced with a fresh one through a port in the bottom of the handgrip.

Borchardt's system impressed a number of manufacturers, but his gun was a little delicate for field use, and the large odd-looking spring housing in the rear made the pistol clumsy to handle. Still, it was successful enough to drive a flurry of competitors into designing similar guns. The Mauser Company developed a comparable recoil-operating automatic pistol in 1896. In some ways, this weapon was less like modern semiautomatic pistols than was the Borchardt; for example, it used the charger/strip-clip system found in Mauser rifles and fixed the magazine in front of the pistol trigger instead of hiding it in hand grip. On the plus side, however, it used a smaller, less intrusive recoil spring and a slightly more powerful cartridge. Like the Borchardt, it too was often used as a light carbine, but the Mauser's clever attachable wooden stock was hollow and designed to do double duty as a holster. The firearm also demonstrated the high level of durability and quality of construction that Mauser firearms were becoming famous for. Winston Churchill purchased one and had it with him both during the Battle of Omdurman and before his capture in the Boer War.

Mauser's weapon was a commercial success and put Ludwig Leowe on the defensive. The firm decided to redesign the Borchardt, a job that was put more or less in the hands of Georg Luger, a former firearms salesman with some talent for engineering. Luger kept a hinged toggle arm and the pistol-grip detachable magazine but changed most other elements of the gun. The Borchardt's magazine grip had been set nearly perpendicular to the barrel, but in Luger's modification the grip was at a sloping angle that allowed for better handling. He also got rid of the awkward rear-spring mechanism and the structure that housed it.

The resulting pistol became one of the world's most famous and recognizable handguns. Officially named the "Parabellum," the gun was initially met with some suspicion on the part of German authorities concerned about the strength of its short-version 7.63mm cartridge. After Luger modified the pistol to use a 9mm nonbottlenecked cartridge, the authorities became more enthusiastic (Hogg 1996, 119–120). The gun was officially adopted by the German army as the Pistole 08 in 1908, but of course became more commonly known as the "Luger."

An inventor as active as John Browning was not about to let the semiautomatic pistol emerge without his input, and in the late 1890s he too began putting together prototypes. Eventually Browning-designed pistols would become as well known and familiar as the Luger. He initially explored the merits of gas operation, but fell back on the blowback system. Starting in 1900 he produced a number of successful semiautomatic handguns, and while each was an improvement on earlier designs they all shared certain similarities. The guns used the detachable magazine system set in the pistol grip like the original Borchardt. The main recoil spring, however, was not behind the bolt but rather alongside the barrel. In his initial Model 1900, it was above the barrel; but in order to improve the aiming features in later versions, it was fixed either below or actually *around* it. A sliding framework (or "slide") that surrounded the outside of both the barrel and the spring further characterized these pistols. Before firing, this slide was manually pulled back toward the shooter. It dragged back the bolt and engaged the rear hammer that was cocked in turn by the motion. When the slide was released, the recoil spring slammed forward and the bolt picked up a cartridge from the magazine that it then drove into the breech. After the initial shot, the force of the propellant operated the bolt and slide so that the entire process happened semiautomatically.

Like many other weapons he designed, Browning's pistols were simple, reliable, and popular. His relationships with American manufacturers like Winchester were usually amiable, but often punctuated by what might today be called, in corporate-speak, "differences in vision." Their loss was another's gain as the Fabrique National d'Armes de Guerre of Belgium, or "FN" as it is more commonly known, proved more compatible with the prodigious gunmaker's sense of innovation. As a result, many of the semiautomatic pistols that Browning designed after the turn of the century were built in Belgium by FN.

One notable and particularly famous exception to that rule came about as a result of the American occupation of the Philippines. In the first decade of the century, U.S. troops were engaged in a conflict against Filipino militants who resisted U.S. imperialism. These insurgents reportedly used narcotics that left them temporarily more resistant (relatively speaking, of course) to bullet wounds. Soldiers in the field complained that it took too many shots from a .38-caliber service revolver to bring down one of these hopped-up fanatics. The Army opted for technology over diplomacy in solving this problem, and they began shopping around for a pistol with more stopping power. Browning teamed up with one of his old American partners, Colt, to submit a design. Perhaps due to its long association with revolvers, Colt had initially shied away from semiautomatic pistols, but now through Browning's considerable assistance they managed to win the army contract with one. The final design, known as the Model 1911, fit the basic pattern of a Browning semiautomatic pistol, but was chambered for a large .45-caliber cartridge. This famous gun would remain the standard U.S. Army service pistol right into the 1980s. For its part, FN continued to turn out a number of different Browning pistols in a variety of other caliber sizes such as .25, 7.65mm, .32, and, perhaps most commonly, 9mm. Other companies followed suit with dozens of similar weapons, some becoming quite famous in their own right such as the Walther PPK carried by the fictional not-so-secret agent James Bond. Most new semiautomatic pistols today, whether made by FN, Colt, Beretta, or Glock, utilize a design not far removed from Browning's originals.

The energetic inventor was still not yet finished developing new guns, however, and would go on to create other successful weapons. Although his pistols used the recoil method, he did not abandon gas operation and sought to employ it in sporting arms. He developed gas-operated rifles and an exceptionally popular five-shot semiautomatic shotgun that used an under-barrel tube magazine, similar to that found in the Henry or Lebel. In the meanwhile, however, others had begun to apply gas operation to new machine gun designs.

THE AIR-COOLED LIGHT MACHINE GUN

An Austrian nobleman, the Captain Baron Adolf Odkolek von Augeza, developed one of the more successful machine guns to compete with Browning and Maxim. Fortunately his gun did not inherit his titles but rather became known as the Hotchkiss after the company that purchased his design. Benjamin Hotchkiss had been an American-born inventor that worked predominantly in France. He had earlier developed several hand-operated machine guns and rapid-firing light cannons, some of which had performed their most dishonorable "service" by helping execute some 200 Lakota civilians in the 1890 massacre at Wounded Knee in South Dakota. The indiscriminate blasts of the Hotchkiss weapons may also have been responsible for the deaths of nearly thirty troopers of the U.S. Seventh Cavalry due to "friendly fire" during the tragedy. Perhaps it is fortunate that Hotchkiss himself missed news of the event, having died five years earlier. The company he founded, however, continued to produce new weapons and toward this end obtained the basic design for Odkolek's improved machine gun. In 1897, the French army adopted the first in a series of these Hotchkiss automatic machine guns.

The weapons had more in common with the potato digger than the Maxim, being air-cooled, gas-operated, and light enough to mount on a tripod. The gun's firing system employed a cylinder fixed directly below and parallel to the barrel. A hole in the barrel itself allowed gas from a shot to flow into the cylinder and drive a piston that in turn moved the breech bolt back. A spring then returned the bolt into place, and it picked up a fresh cartridge as it did so. Although some later versions could use standard ammunition belts, many Hotchkiss designs employed a shorter, inflexible, metallic "strip" of 24–30 cartridges. These machine guns had no water jackets but rather were intended to be air-cooled. While temperature regulation was not perfect, it was partially addressed by metallic fins attached to the barrel to dissipate heat and by the practical limitation of stopping to add a new ammunition strip after every thirty rounds. Overheating was also not much of a problem in a speeding aircraft, and many Hotchkiss machine guns found their way onto planes in World War II. The machine gun was successful enough as a weapon to be adopted for use by a number of nations, in particular the Japanese who valued its portability in their campaigns in rugged East Asia. Versions of the Hotchkiss system were widely used by the French and Japanese Empires right through World War II.

Meanwhile, the Maxim system was also being modified and improved in the early twentieth century. In 1910 the Russians developed a slightly more compact version of it that used a smaller wheeled carriage and thus offered better battlefield flexibility. For its part, the Maxim Gun Company had earlier been purchased by its own chairman, British industrialist Albert Vickers, in 1897. The Vickers Company set about redesigning the gun with an eye to further simplification and weight reduction, and was able to develop a model that was about 25 percent lighter than other contemporary Maxim guns. The modified Russian and British versions of the Maxim gun were incredibly hardy weapons that would see extensive service right through the Korean War.

Browning also continued to produce new machine guns, including a heavy belt-fed weapon that looked a great deal like the Vickers, especially given its large water jacket around the barrel. The construction of the breech mechanism differed from that of the Vickers or Maxim, but like those other weapons this particular Browning was recoil-driven. It was adopted for service the year that the United States entered World War I and thus became known as the M1917. Like its British and Russian counterparts, the basic gun was used by the United States well into the middle of the twentieth century.

Reliable and durable as these water-cooled machine guns were, their use still entailed considerable disadvantages. Resourceful British troops were known to brew tea with the hot water from their Vickers guns, but beyond this subsidiary benefit the water generally made the gun heavy and more cumbersome to deal with. Furthermore, if a water-cooled machine gun ran out of water, it was not of much use. In the desert campaigns of World Wars I and II, careful measures had to be taken to avoid this problem. There are a number of military anecdotes of soldiers running low on water and being forced to urinate on their guns to keep them cool.

In the meantime, the search continued for an air-cooled machine gun that was light and durable. One notable weapon was developed in Denmark by an army captain named W. O. Madsen. The Danes adopted his lightweight machine gun in 1903, and soon other countries such as Russia and Germany began employing it in specialized roles. Eventually the Madsen would be used by thirty-four different countries, appear in dozens of caliber sizes, and, like many other successful automatic machine gun designs, see service well into the twentieth century (Owen 1975, 87–88).

The Madsen's popularity lay in its reliability and low weight of about 20 pounds. It was recoil operated and, although belt-fed and water-cooled versions were produced, the Madsen was generally designed as an air-cooled gun using a detachable banana-clip magazine. Like other light machine guns that would soon be developed, part of the cooling success was achieved through the ammunition feeding system. The largest Madsen magazines only held forty rounds, thereby limiting long periods of sustained fire.

A more innovative air-cooling system appeared on another popular light automatic weapon known as the Lewis gun. It was named after Isaac Lewis, a U.S. Army lieutenant colonel who had been hired to redesign an earlier prototype. Lewis was so pleased with his modified weapon, and so displeased with the Army's lack of interest in it, that he left the service and bought the rights to the gun. He began producing it in Belgium in 1913, but much of the manufacturing was shifted to the Birmingham Small Arms Company of Britain, particularly during World War I after Belgium was largely overrun by Germany. Like the Hotchkiss the gun was gas-operated, but the cooling system was more distinct. It used a large, awkward-looking metal air-cooling jacket or "shroud" fitted around the barrel. The shroud was designed so that the blast from the muzzle created a vacuum effect that sucked in air through a hole near the breech. This created a current that helped keep the barrel cool. The weapon was not belt-fed but rather also used a round drum magazine that held 47 cartridges. Once again, the simple fact that this drum had to be replaced frequently also helped to keep the gun's effective rate of fire and barrel heat low.

In addition to the Lewis gun's great reputation for reliability and accuracy, the weapon weighed less than 30 pounds. This compact size made it ideal for a number of different roles. Unlike many other machine guns, it could easily be carried with a squad or platoon of infantry for mobile fire support. The ability to provide sudden bursts of shots from unexpected quarters reportedly earned it the nickname of "the Belgian Rattlesnake" amongst German troops. These light guns were also easily mounted on a variety of vehicles. One of the important roles in this regard, advocated by Lewis himself, was as an aircraft weapon. In this role it was usually fitted with a larger ammunition drum, and the cooling shroud was removed for obvious reasons.

Not all of these early light machine guns were as successful as the Madsen or the Lewis. The French began producing a similar recoil weapon, known as the Chauchat, during World War I that did not enjoy a good reputation. Besides having a fairly severe recoil kick, it was particularly prone to jamming. Aesthetically speaking, the device had a nice streamlined appearance, although its detachable magazine was perhaps one of the oddest-appearing ever conceived. It looked something like a banana-clip magazine, but curved even more radically to form a 180-degree half-circle. Like the other light machine guns, the Chauchat did have the advantage of being highly portable, although this was more than offset by its poor performance.

Air-cooling systems eventually appeared on most later light, medium, and heavy machine gun designs. Some of the more famous include another Browning design, the Model 1919 machine gun, variations of which are still in service in many countries today. The German army under Adolf Hitler developed a light machine gun with an exceptionally high rate of fire. This MG34 (and later MG42 version) was capable of firing 1,500 shots a minute, but was limited of course by the amount of ammunition that could be carried. The gun helped inspire the American M-60 light machine gun that became famous during the Vietnam War.

The term "light machine gun" is probably best understood, then, as a fairly compact weapon that is portable, like the M-60, but that can utilize fully automatic fire. That definition is not an entirely satisfactory one, however, since a variety of other guns were beginning to emerge as early as World War I that could do the same thing. These were so small, however, that they did not somehow seem to warrant the name "machine gun." Other terminology emerged to describe these firearms, such as "machine pistols," "submachine guns," and eventually "assault rifles."

THE SUBMACHINE GUN

In the search for light automatic weapons, one existing type of firearm could be modified to perform the same type of service with only a little tinkering. The semiautomatic pistol had been inspired by the Maxim gun and operated upon similar principles; therefore, the jump between a semiautomatic pistol and a fully automatic one was not a particularly long one. Modifying the gun's firing mechanism so that it would continue to work through the firing-and-reloading cycle without stopping could be done relatively easily. In fact, fully automatic versions of both the M1911 and Luger were experimented with.

In the latter case, a special Luger pistol had originally been designed for artillery personnel to use for close defense. These were fitted with an 8-inch barrel, a carbine shoulder stock, and a special thirty-two-shot snail-shaped drum magazine that extended out below the handgrip. While this "artillery Luger" was an adequate carbine, it did not suit the bill when converted to fully automatic fire. Like most machine guns, the recoil of one shot tended to spoil the aim of the next. In a heavy, fixed weapon like the Maxim this was less of a problem, but the lightweight experimental Luger machine gun/pistol/carbine tended to spray bullets all over the place. This phenomenon of "muzzle rise" has in fact been an ongoing challenge to designers of automatic weapons, one that has never been completely addressed.

A more satisfactory design was developed in 1916 by Hugo Schmeisser, who was working at the time for Theodor Bergmann Armament Manufacturing. The gun used the same 9mm cartridge and snail magazine as did the artillery Luger, although it was fixed on the left-hand side of the gun, rather than below the pistol grip. It was also designed as two-handed weapon from the ground up with a single-piece complete wooden stock. The automatic gun still bounced around a lot when fired, but the new design allowed for better control. Although it looked more like a carbine than a pistol, its connections to the Luger led to a military designation of

Two British-led Elite Kenyan soldiers. The one on the left is sporting a Sten submachine gun and his colleague has a Lee-Enfield rifle. © Hulton Archive/Getty Images.

"Machine Pistol 18, I" when it was adopted for use by the army in 1918. Although the Germans eventually ordered 50,000 of the weapons, also known as the "Bergmann," the war was reaching its conclusion and less than a fifth of these ever made it into soldiers' hands. One wonders what the outcome of the conflict might have been if the full complement had been delivered prior to Germany's last great offensive in the spring of 1918.

By the time Deutschland had managed to develop a small-caliber, handheld automatic weapon for its troops, one of the Allied countries had also done likewise. The Italian firearm firm of Beretta modified an earlier, more complicated weapon to create a gun that was very similar to the MP18, I. This firearm was known as the "Beretta Moschetto Automastico" or, more formally, as the M1918/30. This gun looked a great deal like its German counterpart, although it used a banana-clip magazine that loaded from the top, reminiscent of a slender version of that employed on the Madsen. Like the Bergmann, the weapon came too late to radically alter the battlefield situation, but together both guns inspired dozens of similar weapons over the following decades. For its part, Beretta developed a series of analogous submachine guns later used in World War II, such as the Beretta Model 1938 A. This continued to employ a wooden stock, but the original banana-clip magazine was straightened and now fixed underneath the barrel. The Bergmann was also overhauled by Germany after the war, and its 1928 version, the MP28 II, likewise sported a straight box magazine. The later Berettas and Bergmanns were also capable of "selective fire," which means that they had the capability of operating automatically or semiautomatically as the shooter saw fit.

When the British and Russians decided to start producing submachine guns, they took a simple route. Rather than designing entirely new weapons from the ground up, they both just made copies of the redoubtable Bergmann MP28 II. The Soviet doppleganger was known as the PPD34, while the British eventually came up with a version was called the Lanchester Mk I. Both of these weapons were named after their respectives "designers," the Russian acronym translating roughly to "machinegun pistol of (Vasily) Degtyarev Model 1934." Degtyarev, incidentally, was a lifelong expert in firearms, having first started working at the Tula arsenal when he was only eleven years old.

The Americans, for their part, came up with a something quite different from the ubiquitous Bergmannesque style. During its brief participation in World War I, the United States had explored several options for portable automatic weapons. Some of these wound up being dead ends, such as a modified 1903 Springfield rifle designed to fire pistol ammunition. A more successful approach was developed privately under the guidance of John Thompson, a retired U.S. Army general. During World War I he put together a team of engineers to develop a lightweight automatic weapon for use in the war. He originally envisioned a larger caliber rifle, but when his designers advised using pistol cartridges Thompson reportedly stated: "Very

Soviet soldiers during World War II sporting the dependable PPSH 41 submachine gun. © Hulton Archive/Getty Images.

well. We shall put aside the rifle for now and instead build a little machine-gun. A one-man, hand held machine gun. A trench broom!" (Helmer 1969, 25). This was not the last nickname Thompson's gun would receive. He recognized that the portable weapon his team went on to build did not fit the popular image of a heavy machine gun. Nor did he see the two-handed, nearly 10-pound gun as a "pistol" in any real sense of the word. He therefore coined the name "submachine" to classify it. This term was subsequently picked up in most English-speaking countries to describe what the Germans continued to call "machine pistols."

Although the Thompson began to be widely manufactured by Colt in 1921, the U.S. Army initially showed little interest in the gun. In those days fully automatic weapons were commercially available, and the new submachine managed to find a civilian "niche market" so to speak. Organizations of illegal alcohol dealers were less wed to convention than was the military, and they began using them in distribution disagreements with each other. As a result, the "tommy gun" or "Chicago typewriter" made a big splash in the U.S. popular imagination. Eventually the army adopted limited numbers of the oft-named gun, now calling it the "M1928." One key difference between this particular American submachine gun and European models was reflective of the distinction between the automatic pistols favored by the different powers. Just as the Bergmann used 9mm Luger ammunition, the Thompson employed the same .45-caliber cartridges utilized by the M1911 semiautomatic pistol. Americans in general felt that the .45 packed a bit more punch than the 9mm, but it is doubtful whether this made much difference to any poor wretch that got caught in a burst of fire from either. Although it was heavier than the Berettas or Bergmanns, the Thompson quickly developed a reputation for being exceptionally rugged and dependable. The look of the Thompson was also somewhat distinct from its Continental cousins. It did not use a single-piece carbine-like wooden stock, but instead had a metal frame with a separate wooden pistol grip and buttstock. Rather than a traditional simple fore stock, there was a second pistol-type grip set closer to the muzzle for the shooter's other hand. The double pistol grips were intended to improve fire control. Between these handholds, under the breech was a port that could accept several different magazines including fifty- and 100-round drums and twenty- and thirty-round straight box magazines. The former in particular were associated with the gangster wars of the 1920s and 1930s, while straight box magazines were more commonly used in World War II. By the time of that conflict, the Thompson's design had been simplified to ease production a bit. For example, the front pistol grip was replaced by a more standard fore stock and cooling rings on the barrel were removed. During the 1920s the gun initially retailed for $200, but changes in design and mass production eventually drove the wartime cost below $50. Although the Thompson had been simplified, it still was a fairly complicated piece of machinery, especially in comparison to new models of submachine guns being produced abroad.

Demands for equipment hit other countries in World War II even more severely, and changes in submachine guns appeared as a result. In Italy, for example, the Beretta Model 1938/42, Model 1938/43, and Model 1938/44 each became progressively more simplified as production pressures increased. The Germans had already been moving in this direction before the war started, and in 1938 they adopted a replacement for the Bergmann. This weapon was designed by Heinrich Vollmer for the Erma company, and operated on the same blowback principle as other submachine guns. Vollmer's gun was quite distinct cosmetically, however. The MP38 and the later MP40 variant were made entirely of metal and used a folding buttstock and pistol grip. The box magazine was positioned forward and extended downwards so as to serve double duty as a grip for the second hand. These guns were widely used by German forces, and the weapon has become as popularly cemented to the image of Wehrmacht troops as the Thompson is to gangsters and GIs.

The exigencies of war forced Britain to come up with a more economical submachine gun too. The Lanchester was a good weapon but took far too much money, time, and effort to construct. American Thompsons were considered even better in combat but also were pricey and often wound up at the bottom of the Atlantic thanks to German U-boats. A cheaper replacement had to be found. The particular submachine gun that came to the rescue was the called the "Sten" and it was developed by two men, Reginald Shepard and Harold Turpin, working at Enfield. Its name was an amalgamation of Shepard, Turpin, and the first two letters of Enfield or (more patriotically) England. It was an exceptionally clever but simple design that could be made quickly and cheaply. The designers initially hoped to be able to produce fifteen Stens for the price of a single Thompson. They certainly exceeded that goal in comparison to the earliest Thompson models, as a wartime Sten only cost £2 and 10 shillings ($10.59) to manufacture. The design was so simple that European underground organizations were able to produce copies in secret using rudimentary blacksmith equipment. For their part, the British cranked out some 4 million Sten guns during the conflict. It was generally considered cheaper to simply buy a new Sten than to waste a technician's valuable time trying to repair a broken one.

There were several versions of the Sten, and some models included wooden stocks and handgrips. The most commonly produced, the Mark II and Mark III, however, were completely metal like the MP38 or MP40. These guns had a simple steel rod with a flat plate to serve as a buttstock. There was no pistol grip, just a bit of extra plating on the rod to grasp. The shooter was expected to use the box magazine as a forward hand grasp, although in the Sten this jutted out from the left side of the gun rather like the old Bergmann. The placement of the left hand made shooting from the hip easy, but lining up an aimed shot with one's eye was a bit awkward. As crude and primitive as these guns were, they operated well enough. In competitive tests held by the U.S. Army, they actually outperformed the beloved Thompson (Helmer 1969, 188–189). As a result of an advanced blowback design that allowed for a lighter bolt, the smaller models weighed only a little over 6 pounds, almost two-thirds that of a Thompson. In the field, however, they developed a reputation for jamming, a flaw the Thompson was rarely accused of.

By the time Germany attacked Russia in World War II, the Soviets had likewise redesigned the original PPD 34 into a similar PPD 40 that now sported a two-piece wooden stock. The separation of the fore stock from the butt was made so that the gun could more easily accept a large seventy-one-round cartridge drum. Like other countries' Bergmann-based submachine guns, however, the PPD 40 was also surpassed by something even easier to manufacture. The urgency in the Russian case was particularly severe because the Germans had captured many of the production facilities in which PPD

40s were made. The new model was called the PPSH 41, designed by Georgii Shpagin. Like other cheaply made submachine guns of World War II, it used a lot of simple stamped metal parts. It retained a wooden buttstock, however, presumably because wood remained a bit more economical in the vast Soviet Union than it was in clear-cut England or Germany. The fore stock, however, was dispensed with, and the shooter was expected to either grasp the cartridge drum or the area directly in front of the trigger. The barrel, interestingly enough, was often made of old bolt-action rifle barrels that were chopped in half to form two submachine gun barrels (Barker and Walter 1971, 31). Eventually the Russians besieged in Leningrad developed an even cheaper, all-metal submachine gun called the PPS 43 that looked a little like the German MP40. Although this proved to be another clever and reliable design, it did not replace the PPSH 41, which was cranked out by the millions. Just as German troops of World War II are frequently associated with the MP40, the popular image of a Soviet soldier is inevitably holding a PPSH 41 fixed with a round drum magazine.

Although it was wealthier than its allies, the U.S. military was not immune to budgetary concerns, and the army began looking for a more cost-effective supplement for the Thompson. Its $20 solution was a weapon known as the M3 submachine gun. This gun had been designed in one month by George Hyde of Inland Manufacturing and was adopted for service at the end of 1942. Like other mass-produced machine guns, it was all metal and used stamped parts. The design looked very much like the Russian PPS 43 or the German MP40, with a pistol handle for one hand and a straight box magazine doubling as a handgrip for the other. It too had a retractable metal buttstock, but continued to use the heavier .45-caliber cartridge just like the Thompson. It also had a distinctively cylindrical shape that reminded troops of a mechanic's lubrication device, and the M3 was subsequently nicknamed the "grease gun." Although it too had some reputation for jamming, it was reliable enough to remain in U.S. service through the 1950s. The ultimate reason for its eventual disappearance had less to do with performance than with a general decline in the role of submachine guns.

Besides the really famous submachine guns noted above, there were many other designs used during and after World War II. Some of these became quite noted in their own right. The German Heckler and Koch MP5, the Israeli Uzi, and the U.S. Ingram Mac 10 and Mac 11 were all famous submachine guns of the Cold War. Nevertheless, as successful as these compact automatic weapons had been, they would not predominate on the battlefield for long. In spite of their television and cinema fame, in the second half of the century submachine guns like these were increasing resigned to use by specialized military personnel, police, and more contemporary distributors of controlled substances. Most regular line soldiers were not issued guns like these. The situation had actually been the same for the majority of troops fighting during World War II. As noted above, the popular image of a soldier in this conflict almost inevitably carries a submachine gun, but the fact of the matter is that the average private was more likely to be dragging around an older bolt-action rifle. These weapons, be they Mausers or Lee-Enfields, were little changed from the late nineteenth century. Rifles continued to offer advantages that submachine guns failed to deliver.

Submachine guns used pistol ammunition, which lacked long-range accuracy and hitting power. The reason for this can be a little confusing, as a 9mm pistol cartridge sounds like it should be more powerful than a 7.62mm rifle cartridge. Keep in mind, however, that bolt-action rifles used *bottlenecked* cartridges. Because of the extra propellant in the rear of the casing, the projectile received relatively more push when the propellant ignited.

A collection of automatic weapons. Clockwise from upper-left-hand corner: Italian Beretta Model 1938 A, U.S. M-2 Carbine, U.S. late-model Thompson submachine gun, German MP40, German StG 44, Soviet PPS 43, French MAS 38, U.S. early-model Thompson submachine gun. © Hulton Archive/Getty Images.

The standard pistol cartridge is "straight-walled" but short, leaving relatively less room for propellant. Bottlenecked rifle cartridges are simply too long to fit in the narrow magazine typically used in a semiautomatic pistol or fully automatic submachine gun. Lastly, of course, the rifle's higher muzzle velocity is also aided by the simple fact that it has a longer barrel that is able to take better advantage of the expanding gases. In the end, a typical .45 pistol cartridge used in a M1911 will have a muzzle velocity of around 850 feet per second. A thinner, but longer 5.56mm (.223-inch) cartridge fired from a modern M-16A2 assault rifle has a muzzle velocity of about 2,800 feet per second. It is true that the .45 bullet weighs a bit more, but the additional mass is not enough to deliver more damage than that caused by the smaller round traveling three times the speed. The higher velocity also allows for a flatter trajectory and a longer range. For all of these reasons, bolt-action rifles remained popular through World War II despite their relatively low rate of fire.

THE ASSAULT RIFLE

During World War II, a historian of firearms considered the immediate future of the basic infantryman's armaments. Dismissing the submachine gun as an auxiliary device, he pondered what type of weapons would remain most useful on the battlefield. "For the new mobile infantryman who follows the tanks, storms the strong points, aims at the hostile light-armored vehicles, machine guns, anti-tank guns, airplanes, it looks as though his weapons in the next few years would be whittled down to the (1) semi-auto rifle . . . (2) light mortar . . . (3) automatic rifle or light machine gun" (Newman 1942, 60). The writer turned out to be quite prophetic. Although it took more on the order of several decades, the common infantryman did wind up with those three types of weapons. They were, however, combined into a single weapons system. This new gun could be used as either a semiautomatic rifle or a light machine gun. When fitted with an attachable grenade launcher, it could even copy the function of the small mortar envisioned above.

The weapon causing the demise of both the long-range bolt-action rifle and the speedy submachine gun was the automatic or "assault" rifle that came to incorporate many of the best features of both. The term "assault rifle" generally refers to a handheld firearm capable of selective fire, using bottlenecked cartridges stored in a 20–30-round box magazine. The difference between the way a semiautomatic rifle and an assault rifle operates is fairly narrow, similar to the relationship between a semiautomatic pistol and a submachine gun. Like so many firearms throughout history, the so-called assault rifle was anticipated as a weapon well before practical models emerged.

Even before World War I, a number of prototype rifles capable of automatic fire had emerged. One example was the Griffiths and Woodgate Automatic Rifle, which operated upon the recoil principle used in the Maxim. This weapon never gained widespread support, probably because of excessive muzzle rise. During World War I itself, Browning came up with something similar that enjoyed much more success. He had developed gas-operated semiautomatic sporting guns, so it was a relatively simple matter for someone of his abilities to come up with a similar automatic weapon. The end result was the famous Browning Automatic Rifle or "BAR." This gas-operated gun was developed in 1917 and saw wide service with the U.S. army and many other nations through the 1950s. It had a twenty-round detachable box magazine and fired the same .30-caliber (7.62mm) cartridge used in the 1903 Springfield. In a way, the gun was a bit of a hybrid. The fact that Browning himself called it an automatic rifle suggests that it was not quite a light machine gun, but something else. Since it weighed about 19 pounds, it was exceptionally light for a machine gun but was only about twice as heavy as a typical rifle. Its portability is attested to by the fact that no less than three of these guns were found in Bonnie and Clyde's car trunk, after that psychotic couple was mowed down by a team of quasi-legal assassins, one of whom had brought his own BAR. Although early models had a single-shot setting, later versions were limited to two rates of automatic fire and came equipped with a bipod stand. Generally the weapon wound up being used as a more of a light machine gun than anything else.

The utility of semiautomatic fire in a basic infantry rifle was recognized by the United States, which began looking for something of the sort to replace the 1903 Springfield with. By 1936 they had adopted a semiautomatic gas-operated rifle designed by Canadian-born John Garand, an employee of the Springfield Armory. Known as the M-1 "Garand" rifle, this firearm used eight .30-caliber (7.62mm) cartridges held together in something called an "en bloc" clip. This clip was loaded into the gun through the top of the breech, and its particular design did not permit the magazine to be topped off once several shots were fired. The lack of a "top-off" feature was a minimal handicap at best when one considers the amount of semiautomatic fire this weapon could throw out in comparison to a bolt-action rifle. During World War II some of the weaponry of the U.S. Army lagged behind that of its enemies, the easily ignitable Sherman tank being perhaps the prime example. The same criticism cannot be made

U.S. soldier during the early Cold War aims an M-1 (Garand) rifle. © Hulton Archive/Getty Images.

of the semiautomatic Garand, which was the best standard line infantry rifle in the world. The Garand won high tribute from the common grunt to the loftiest generals with Dwight Eisenhower, Douglas MacArthur, and George Patton all voicing their praise. An American rifleman could tear off three shots for every one fired by his British, German, Italian, Japanese, or Russian counterparts. In spite of its technical superiority, the rifle was rugged and could operate well in conditions of mud or foul weather. It is also true that while in combat, American troops would sometimes take it upon themselves to tinker with their Garands so that they fired as a fully automatic weapon. The muzzle rise caused by such a heavy cartridge in a relatively light rifle, however, had to have been significant.

Although the Garand was a great success like the BAR, it too cannot be considered the direct ancestor of today's assault rifle; rather, that distinction goes to a German weapon developed late in World War II. The German Wehrmacht experimented with a number of handheld automatic weapons besides the MP38 and MP40, but the one that deserves special attention was known variously as the MP43, MP44, or StG44. It was a revolutionary firearm that most assault rifles in use throughout the world today can trace their lineage to. The gun was developed by a team working under the direction of Hugo Schmeisser at the C. G. Haenel company. It used a special "kurz" or "short" version of the venerable Mauser 7.92mm cartridge. This cartridge had the same diameter as the regular Mauser ammunition, but used less propellant. The Germans created it after recognizing that the typical service rifle had an effective range well beyond the distance at which most combat actually took place. This shorter version of the cartridge offered a compromise between the accuracy of a more powerful traditional rifle projectile and the automatic-fire potential of lower powered pistol ammunition.

The gas-operated StG44 made good use of the new ammunition, having the ability to function at normal combat ranges with either semiautomatic or fully automatic fire. Like other mass-produced weapons of World War II, much of the gun was made out of stamped metal parts, although it did have a wooden buttstock. It also had a pistol grip, but like the Henry rifle there was no fore stock. Cartridges were fed into the breech from a thirty-round banana-clip magazine fixed underneath the barrel, directly in front of the trigger. Its various name changes came about for several reasons; initially the "machine pistol" designation was made to hide the project from Hitler, who had forbidden further work on experimental rifles of these types. The designation was not necessarily a bad one in any event, since the firearm could perform a similar task to that of MP40 anyway. Once good reports about the gun began flooding in from the field, it earned the more dramatic name of "Sturmgewehr," hence the "StG" designation. The term, which may have been coined by Hitler himself, is generally translated into English as "assault rifle," although "storm rifle" would probably be more precise. For whatever reason, the former translation stuck and has come to be used to describe an entire class of firearms.

The gun quickly gained a strong reputation, and German units in the field tried to get as many as they could. Like other Nazi wonder weapons, this advanced rifle was developed too late and although tens of thousands were being cranked out by the end of 1944, most German soldiers never had the chance to use one. Still, enough of the weapons were distributed to troops on the eastern front for some of them to make it into the hands of the Russians. They would go on to inspire a similar weapon that some might call the world's most successful assault rifle, and others might simply call the world's most successful firearm.

The job of designing a Soviet version of the storm rifle was given to an ex–tank commander, Mikhail Timofeevich Kalashnikov, who had taught himself to design firearms while recovering from a shoulder wound received early in the war. His work lasted past the end of the fighting, but by 1947 his new assault rifle designated the "Automatic Kalashnikov" was ready for an initial series of tests. After two years of this and some further modifications, the "AK47" was adopted for service. Like the StG, the weapon was gas-operated and had selective fire capabilities. Superficially the gun also looked much like its German inspiration with a pistol grip, wooden stock, and thirty-round banana-clip magazine. It did, however, use a slightly different short 7.62mm (.30-caliber) rifle cartridge and had a wooden fore stock that aided aiming and handling.

The AK47 is a legendarily tough weapon that has held up in terrible field conditions ranging from Siberian tundra to Southeast Asian jungle, Middle Eastern desert, and African savanna. This durability, combined with the weapon's portability, solid accuracy, and ability to lay down heavy automatic fire when needed, made the gun a worldwide favorite in the second half of the twentieth century. In particular, these features endeared the weapon to insurgents who in many cases were fighting against western-allied regimes during the Cold War. The rifle itself actually became a leftist symbol of liberation and resistance, one that was at least as popular and recognizable as the sickle and hammer, the red star, Chairman Mao's *Little Red Book*, or Che Guevara's scraggly mug. The guns went a long way toward restoring a global military balance in firearms that had been drastically offset by bolt-action rifles and the Maxim machine guns in the previous century. Official and unofficial copies of the rifle were made in Soviet client states like China and throughout Eastern Europe. The USSR, for its part, brought out a plethora of different models and upgrades over the years. These include very similar weapons with folding stocks for paratrooper service, short barrels for submachine gun–type duties, or longer barrels, bipods, and drum magazines to operate as light machine guns. Estimates of the number of Kalashnikov-type weapons that have been produced worldwide range anywhere from 30 million to 100 million, making it the world's most popular firearm. There are so many of these weapons floating around today that one can easily be obtained across the nonindustrialized world for relatively little money. One study estimated the worldwide average price of the AK47 in 2002 to be only U.S. $135 (Batchelor and Krause 2003, 86). As far as money and profits go, because Kalashnikov was working for a salary at a state-owned industry in a nation that did not closely adhere to international copyright laws, he made next to nothing for his invention. More recently, he has tried to use his famous name to market a different Russian product that is far less deadly: vodka.

Like so many earlier types of firearms, once the basic technology of the assault rifle was established, there was relatively little to prevent anyone

Afghan anti-Soviet insurgents armed with Soviet-designed AK47 rifles. © Hulton Archive/Getty Images.

from developing their own variation. Nevertheless the Soviets' great rival, the United States, dithered around with different designs for nearly two decades before finding a rival to the Kalashnikov. As effective as the Garand was in the early 1940s, it had clearly been outclassed by the following decade, yet the United States remained unwilling or unable to adopt a similar gun. One candidate was a small weapon developed in World War II known as an M-2 carbine. This lightweight gun was capable of selective fire and used a .30-caliber (7.62mm) short cartridge loaded in either 15- or 30-round magazines. The firearm functioned well, but because it was originally designed for use by rear-echelon personnel, the army decided it needed something more powerful for frontline troops. One branch of the American military establishment still seemed to think that a rifle capable of blowing a Mastodon into tiny chunks at a thousand yards remained a necessity. In general, it appears that what was really hoped for amongst many was a Garand with select fire. One retired serviceman wrote about the older rifle in 1957 and anticipated that "an improved version of this rifle will appear in due time and will have a selector so the operator will have a choice of either semi-automatic or full automatic fire" (Hicks 1957, 117). That same year saw the fulfillment of such intentions in the disastrous M-14 rifle program. The gun had been adopted after years of research and development costing tens of millions of dollars, but was little different from the rifle that it was supposed to replace. This brutal-looking behemoth (imagine a big nasty Garand with a large external-projecting box magazine) fired a slightly smaller version of the .30-caliber (7.62mm) cartridge, but weighed over 11 pounds. It was supposed to use selective fire, but the powerful ammunition made it highly uncontrollable in automatic mode. Some shooters testing the gun in this manner reportedly received bloody noses as a result. Most of the guns wound up being modified to fire only in semiautomatic mode, which was exactly what its predecessor did in the first place. The powerful M-14 actually made an outstanding sniper rifle when fitted with a telescopic gun sight, but it was not properly suited to the new requirements of a general infantry weapon. Although this leviathan enjoyed a measure of popularity with the troops, particularly the Marine Corps, its production was abruptly cancelled in the early 1960s.

The weapon that the United States adopted to replace the M-14 was every bit as controversial as its behemoth predecessor. Within the military establishment, there was another faction who advocated for a smaller caliber weapon more like the AK47. In 1959 Colt purchased the manufacturing and marketing rights to a similar gun they thought showed promise. The firearm was known as the AR-15, and it was largely the creation of Eugene Stoner, a designer who worked for Aramalite. Colt then began using its considerable influence to try and get the U.S. Army to adopt it, but encountered much resistance. During one notorious test, the army put the M-14, the AR-15, and the AK47 up against each other. The M-14 performed better than the lighter Colt, but in fact the parameters of the test were based upon performance specifications developed around the Garand. Since the M-14 was little more than a Garand anyway, it obviously beat out the AR-15. The AK47's performance in these tests was kept secret, however, probably because it won (Hallahan, 1994, 483–485). The Air Force purchased the smaller rifles, however, and general interest in the weapon began to slowly rise. The gun was also aided by the support of the yet-unsullied secretary of defense, Robert McNamara.

This weapon, given the military designation M-16, was to become another classic assault rifle. Although its select-fire operation was not unlike that of the AK47, its appearance was quite different. It was thinner looking and had high raised gun sights, the rear of which doubled as a carrying handle. The whole thing was black in color and made entirely of metal and plastic. Its unique appearance gave the gun a somewhat futuristic look that probably aided its popularity in the early days of the space race, and impressed best and brightest gee-whizzers like McNamara.

U.S. soldier with an M-16 assault rifle fitted with a grenade launcher. Courtesy of Painet Inc.

It also fired a .223-caliber (5.56mm) cartridge that was even smaller in diameter than the one used in the AK47, although it was just a touch longer.

Any rifle that was so radically different from its immediate predecessor was likely to be received with some skepticism by those familiar with the previous model. In the hands of a veteran accustomed to the M-14, Stoner's lightweight, partially plastic rifle must have seemed like a toy. The gun also had some initial troubles because of a low-quality propellant used in the ammunition that caused excessive fouling. Unless carefully cleaned, the weapon was highly susceptible to jamming. The gun quickly picked up a bad reputation, and stories circulated about parents of servicemen mailing gun-cleaning kits to their sons and Viet Cong guerillas not bothering to pick up abandoned M-16s. Another persistent rumor arose that claimed the lightweight semiplastic gun was actually made by the toy manufacturer Mattel. Exaggerated as some of this folklore may have been, the gun clearly was not anywhere near as forgiving of neglect as was the durable AK47. By the late 1960s, these initial problems were being addressed and the army had adopted the weapon as its general service rifle. The latest version of the weapon, the M-16A2, still holds this position. Ironically, the latest variant contains a device that limits its automatic fire to short three-round bursts. This modification is a testament to the fact that it remains extremely difficult to control automatic fire, even with a relatively light-caliber firearm.

Many other countries came up with other versions of the assault rifle. Initially these usually employed the same heavy cartridge as the M-14 so that the United States and its NATO allies could share ammunition. These weapons included the Belgain FN FAL, the German G3, the Italian BM59, the Japanese Model 64, and the Spanish CETME. Although the Swiss were not members of NATO, their SIG 510-4 rifle was designed to use the same standard cartridge in part to aid export sales. Appearance-wise, all these other assault rifles were laid out along very similar lines except the Italian gun, which looked much like the M-14. The others used pistol grips, a forward bottom-fixed box magazine, and nonelevated aiming sights. Most of them also had no wooden components. After the United States shifted to the smaller 5.56mm cartridge, a number of other countries also followed suit. The newer Belgian FN CAL and German HK 33 were each very similar to the older FN FAL and G3 respectively, but used lighter cartridges. The Italian Model 70/223 and Swiss SIG 530 are likewise very similar smaller caliber rifles. The Israeli Galil rifle fires a light 5.56mm cartridge like the M-16, but owes much of its design to AK47s captured during the Six-Day War (Owen 1975, 61).

One of the most interesting and innovative assault rifles to emerge after World War II was developed by the British at (where else?) Enfield. This revolutionary design was created in 1949 by Edward Kent-Lemon and Stefan Jansen. The former was a lieutenant colonel, while the later was an immigrant who had fled from his native Poland during the war. Officially known as the EM 2, it was more popularly referred to as the "Bullpup." It cleverly placed the magazine near the buttstock, *behind* the pistol grip. This shifted the weight of the weapon to the rear, allowing the muzzle to be maneuvered around more easily. Because the bolt traveled back into the stock, it had to be made completely straight. Such a design would normally make aiming tricky or impossible, but the EM 2 pioneered the combination raised sight-carrying handle later copied by the M-16 for similar reasons. Although the design was quite clever, its intermediate .280-caliber cartridge clashed with NATO plans for a larger standardized ammunition, and the gun never entered service. It did inspire later Bullpup weapons like the rugged FAMAS, which was developed by France in the 1970s.

Britain eventually returned to the Bullpup design in 1985 with the dubious SA 80. This lighter caliber remake of the EM 2 is exceptionally accurate, but has been plagued by a number of service problems in the field, particularly in the gritty conditions of the two Gulf Wars. It remains Britain's standard service rifle today, but may not be for long if these problems continue to occur. More promise is being shown by Israel's newest assault rifle, expected to replace the Galil. The TAR-21 (for Tavor Assault Rifle—twenty-first century) program is based upon the Bullpup design, but is quite a bit shorter than a traditional assault rifle. It measures a mere 28 inches in length, 11 inches shorter than an M-16A2 assault rifle, and even 1 inch shorter than the petite Sten Mk II. An assault rifle of this compact size certainly seems to render the submachine gun somewhat redundant. As radical as the Israeli gun may look visually, ironically, its function and application may not be all that far removed from the original Nazi weapon that gave rise to the assault rifle. So the question perhaps arises: where does the firearm go from here?

8

Conclusion: The Limits of Lethality?

◆

So where does the firearm really stand today? Is it still the centerpiece of military operations, or has it been reduced to an action movie accessory? It is clear that the assault rifle, for all its glamour, does not dominate the modern battlefield to the degree that the more primitive Maxim gun and bolt-action rifle controlled the campaigns of World War I. It can hardly be claimed that the outcome of either of the two Persian Gulf Wars has had much to do with the relative merits of either the M-16 or the AK47. Those wars were dominated by airpower, night-vision equipment, stealth technology, mobility, and a variety of other factors. Everything else being equal, the campaigns might have been largely the same even if allied co-alition forces had been carrying World War II–era Garand rifles. On the part of the Iraqi opposition forces, most of the casualties they have in-flicted upon U.S. troops in the second conflict have been caused by crude explosive devices, not through firefights or sniper attacks. It is true that the venerable AK47 continues to play a major role in many regional con-flicts throughout the world, but it is also apparent that the majority of these conflicts are not being conducted with cutting-edge military tech-nology. For that matter, the experience of American and Soviet forces in Vietnam and Afghanistan leaves one to wonder how important technol-ogy really is in certain cases.

Recognizing that the assault rifle may not be the most important component of warfare today, one must wonder about which direction the firearm will move in. There have been a few innovations that suggest future possibilities. One of the most interesting was a weapon the Germans tested in the 1980s known as the G11. This gun actually uses *caseless* ammunition that has the propellant and ignition material packed into the rear of the bullet itself, similar to the original Rocket Ball idea of the 1850s. Despite some initial problems with hot barrels inadvertently igniting or "cooking off" shots prematurely, the final version of the gun performed well in field trials. It was ultimately not widely adopted by the German army due to budgetary reasons.

For its part, the American army is trying to revisit the ancient issue of accuracy in order to improve the firearm's potency. A variety of high-tech optical scopes with range-finders and other features have been experimented with toward this end. One American firearm under development, the XM29, now seeks to increase the weapon's power by including an option to fire explosive ammunition that does not actually have to hit the enemy in order to inflict casualties. Assault rifles are often fitted with an attachable grenade launcher, but the XM29 program is an attempt to more closely integrate both systems together. The weapon is intended to be able to fire standard 5.56mm cartridges and 20mm high-explosive air-burst munitions. The idea is to be able to hit an enemy hiding behind a structure or in cover with a fragmentation blast.

At the dawn of a new century, the firearm's immediate future is not clear. Caseless ammunition and exploding bullets sound innovative, but both ideas were experimented with in the nineteenth century. If we narrowly define old age as simple obsolescence, than guns clearly retain some virility. On the other hand, if one chooses to equate seniority with a lessening of potential and a gradual slowing in the *rate* of development, then the firearm may well be in its golden years. The twentieth century, taken as a whole, clearly did not see the kind of technological leaps in weaponry that were witnessed by the nineteenth. At the dawn of that period, the flintlock was the predominate weapon. One hundred years later, metal cartridges and automatic fire were well known. What did the next ten decades then bring? The century was one more geared toward refining and perfecting earlier designs, rather than inventing anything exceptionally new.

It seems unlikely at the present moment that any particularly groundbreaking designs will emerge based upon the existing materials and technologies employed in current firearm designs. The idea of blasting a projectile down a tube may stay around for a while, but we will need a radically different chemical propellant or even *type* of reaction if the basic potential of the firearm is to be significantly changed. While many new firearms have a highly "sci-fi" appearance, they are not all that intrinsically different from weapons used much earlier. Whatever future handheld weapons *do* emerge, it is difficult to imagine them without some version of a gunlock, a barrel of sorts, and a hand-grip or stock used to hold the thing. In the end, Pandora's laser pistol may not look that much different from her Colt revolver.

At the moment, however, it is hard to imagine the sudden appearance of anything dramatically futuristic like a ray gun, a phaser, or an explosive bolter. On the other hand, the flintlock system had remained predominate for well over a century before Forsyth tried to weatherproof his fowling piece. This kind of breakthrough made by a Forsyth, or a Vielle, or a Maxim has happened in the past and could occur again. A period of great firearms innovation must certainly have seemed less likely in 1800 than it did in 2000.

Furthermore, there is little likelihood that the underlying motivation behind the rise of firearms will be purged as an element of human psychology anytime soon, if ever. Ultimately, it is the same impulse that led the medieval Chinese to stuff bits of shot into a firelance that lies behind the air-bursting XM29. The ancient imperatives of accuracy, power, and rate of fire remain as compelling as ever. As long as people retain their desire to inflict harm at a distance, some version of the firearm will likely be around to perform that notorious service.

Glossary

Arquebus. An early modern firearm that was traditionally lightweight enough to be used by one person and fitted with a matchlock ignition mechanism.

Assault rifle. A somewhat vague term that generally refers to a relatively lightweight one-person rifle with selective fire capability.

Automatic. Automatic weapons are guns capable of firing a continuous series of shots by merely keeping the trigger mechanism depressed after the initial discharge. The first shot will typically require the gun to be manually cocked, but it will then fire in the manner so indicated until the magazine is empty.

Ball. A somewhat dated term for a bullet. It began to decline after cylindro-conoidal and other nonrounded projectiles were developed.

Barrel. Perhaps the key component of a firearm, this is a metal tube designed to restrict expanding gas and help direct a projectile in a particular direction. It may be rifled or smoothbore.

Black powder. The oldest formula of gunpowder, made up of a mixture of saltpeter, sulfur, and charcoal. Its dark color in comparison to some early smokeless powders led to this name.

Bolt-action. A type of firearm reloading mechanism that is manually operated by a small knob.

Bore. The cavity running through the length of the barrel.

Bore weight. An older method of calculating the bore size of a barrel using projectile weight rather than diameter. The figure was calculated by determining how many of the gun's bullets would equal a pound in weight. This system is still used for calculating shotgun bore size, commonly referred to as "gauges" in this case.

Bottleneck cartridge. A cartridge with a propellant case that is of a larger diameter than the bullet.

Bow weapons. Weapons that fire an arrow or bolt projectile by means of released tension from a wooden, metal, or composite bow.

Break-open. A type of breechloading firearm that utilizes a hinge and latch system. Engaging the latch allows the barrel and fore stock to swing downward upon the hinge device, thus exposing the breech for loading and reloading purposes. Also sometimes called hinge-action.

Breech. The "rear" part of the barrel, near the shooter.

Breechblock. The part of the gun that seals off the rear of the barrel. This might be a fixed section as in a musket, or a moveable component as in a breechloader.

Breechloader. A firearm that is loaded through an opening in the breech, as opposed to a muzzleloader.

Bullet. A projectile that is fired from a gun. Typically it is made out of lead with a metallic casing, although other materials such as stone or iron have been used in the past. It is one component of a cartridge. Firearm enthusiasts often claim guns don't kill people, people do. It is more appropriate to say that guns don't kill people, *bullets* do.

Buttstock. The rearmost part of the stock. The section that butts up against the shooter's shoulder.

Caliber. A method of expressing the size of a firearm barrel using a measurement of the bore diameter. Traditionally measured in fractions of an inch in the United States and Commonwealth countries, but also frequently expressed in metric millimeters.

Caliver. An early modern firearm. It was a larger weapon than an arquebus, but lighter than the original musket. Its name may be linked to the word "caliber."

Cap-and-ball. Refers to a percussion cap and lead bullet. The term was often applied to those revolvers that predated fully metallic cartridges, such as the early-model Colt weapons.

Carbine. A shorter version of a musket or rifle, designed for cavalry or other specialized service in which a long firearm might prove cumbersome.

Cartridge. An ammunition packaging device developed in the early modern period. It was originally made of paper and included a projectile (or projectiles) and gunpowder. Eventually came to be constructed of metal and to incorporate a percussion ignition mechanism.

Center-fire. A type of ammunition with a percussion primer fixed in the center of the rear of a metallic cartridge.

Clip. A device used to "charge" or fill a magazine. Often used popularly and somewhat imprecisely to describe a detachable magazine.

Cock. In modern times, this refers to the action of preparing a gun to fire by either pulling back a hammer or firing pin mechanism. Originally referred to the beaklike device that held a piece of flint in a flintlock weapon.

Co-viative. A term to describe the expulsion of early projectiles that do not closely match the shape or size of a weapon's bore. Much of the force of the expanding gases slip past the projectile; similar to the idea of windage.

Deflagration. A fast-burning or explosion with a shock wave that travels slower than the speed of sound.

Detonation. An explosion with a shock wave that travels faster than the speed of sound.

Doghead. The canine-looking device that held a piece of flint on a wheel lock weapon and on certain types of flintlocks.

Firelance. A pole-mounted tube weapon developed in medieval China to operate as an early flamethrower. The immediate predecessor of the firearm.

Firing pin. A pointed device that is driven into the percussion primer in the rear of a metallic cartridge in order to ignite its propellant.

Flintlock. A type of gunlock that used flint and steel in order to set off a powder charge, it dominated firearm designs from the seventeenth through the nineteenth centuries.

Fore stock. The forward part of a gunstock fixed under the barrel. It allows the nontrigger hand to support the often-hot barrel without touching it. This important device was sometimes dispensed with in hastily manufactured weapons.

Gas-operated. A type of automatic and semiautomatic firing system. It uses some of the power of the expanding gases from the gun's discharge to drive an automated reloading system.

Grain. Can refer to granules of gunpowder or a measurement of weight. In the latter case, a grain is 1/7000 of a pound. This might be applied to propellant or to the projectile, such as a 380-grain bullet.

Gunpowder. A dry flammable material that creates an enormous amount of gas when ignited. This reaction allows it to serve as a propellant in a firearm.

Hammer. This component of a gun's lock was originally developed to strike a percussion cap and so ignite the cartridge's propellant. In later weapons, it might serve to strike a metallic cartridge's primer or to drive a firing pin.

Handcannon. The oldest firearm consisting of little more than a metal tube with a touchhole, typically it was fixed to a staff or tiller.

Hinge-action. A type of breechloading firearm that utilizes a hinge-and-latch system. Engaging the latch allows the barrel and fore stock to swing downward upon the hinge device, thus exposing the breech for loading and reloading purposes. Also known as the break-open system.

Lever-action. A type of firearm-reloading mechanism that is manually operated by a lever. The lever was often incorporated into the trigger guard.

Lock. The part of the firearm that encompasses the ignition system; probably descended from the locking mechanism on a crossbow.

Lockplate. A flat metal plate set in the side of the wooden stock near the breech of the barrel. Many of the key components of the lock were fixed to this structure.

Machine gun. A heavy weapon capable of automatic fire.

Machine pistol. A two-handed firearm capable of automatic fire, typically designed to use lower powered pistol ammunition. Another term for submachine gun.

Magazine. A part of a firearm that holds cartridges. This might be a fixed component of the gun or a detachable container.

Match. A cord treated with saltpeter that burned slowly.

Matchlock. An early modern ignition system that used a lit match to set off a powder charge.

Mini-ball. A cylindrical bullet with a rounded top and a hollowed-out base. The cavity in the base allowed the sides of the projectile to swell outward and grip the rifling as it traveled down the barrel.

Musket. Originally a massive weapon of the early modern age. It grew lighter over time and eventually became an almost universal world firearm from circa 1680 to 1850. These guns were invariably smoothbore muzzleloaders.

Muzzle. The "front" end of the barrel, from which the projectile emerges.

Muzzle velocity. The speed at which a projectile is traveling when it just leaves the gun. Usually expressed in terms of feet per second.

Muzzleloader. A firearm that is loaded through the muzzle, as opposed to a breechloader.

Percussion. Nineteenth-century ignition system that used a percussion cap containing fulminate. When struck with a hammer, this material ignited and set off the powder charge.

Pistol. A small gun generally designed to be fired with one hand.

Primer. This originally referred to the material that is first ignited in order to set off the main charge. Later came to mean a percussion cap or the particular part of a cartridge that is struck for the same purpose.

Pump-action. A type of firearm-reloading mechanism that is manually operated by moving a slide mechanism backward and forward. This slide is usually attached to the fore stock. Originally known as a slide action.

Ramrod. A narrow rod made of wood or steel, used to push a projectile down the barrel of a muzzleloader.

Recoil-operated. A type of automatic and semiautomatic firing system. It uses the power of the recoil from the gun's discharge to drive the reloading system.

Rifle. A type of firearm with a rifled barrel. A rifled barrel has shallow spiral grooves cut in its inside in order to provide spin to a fired bullet.

Rifled musket. An infantry weapon that combined elements of a rifle with features of a musket. It had a rifled barrel, but was loaded through the muzzle. Its development was dependent upon the mini-ball.

Rimfire. A type of ammunition with its percussion priming material set along the rear rim of a metallic cartridge.

Round. A term that is often used to describe a cartridge. The term was probably originally linked to the lead "round ball" ammunition fired from a musket.

Selective fire. This is a feature that allows certain weapons to fire in different manners. Typically this will consist of automatic and semiautomatic settings, although some machine guns have been able to use different rates of automatic fire.

Semiautomatic. Weapons that have this capability can fire multiple shots by repeatedly pulling the trigger. The initial shot will typically require the gun to be manually cocked, but it will then fire in the manner so indicated until the magazine is empty.

Serpentine. A component of a matchlock. The upper end of the serpentine held a smoldering match that was used to ignite the powder charge.

Sights. Attachments that are used to help aim a gun. They usually consist of a front sight fixed near the muzzle and a rear sight near the breech. By carefully lining up both sights with a target, the shooter increases her chance of hitting it. Some firearms utilize a telescopic variation.

Smokeless powder. These propellants were first developed in the nineteenth century based upon nitrocellulose. Technically speaking they do produce smoke, but not to the degree of older black powders.

Smoothbore. This can refer either to a barrel with no rifling or to a gun fitted with such a barrel.

Snap-lock. An early version of the flintlock firing mechanism. Like the flintlock, it too used the principle of flint striking steel to set off a powder charge.

Stock. The part of a firearm that is designed to be held and handled by the shooter. In more traditional firearms, the stock was often made from a single piece of wood, but in later weapons it might be composed of synthetic materials and be of a variety of designs. It often consists of a separate buttstock and fore stock.

Submachine gun. A two-handed firearm capable of automatic fire. Typically designed to use lower powered pistol ammunition. Sometimes called a machine pistol.

Tiller. The part of the crossbow that is grasped by the archer and that the bow is set perpendicular to. An important predecessor of the gunstock.

Touchhole. A hole drilled through the breech of the barrel through which an ignition travels in order to set off the main charge.

Trajectory. The path on which a bullet or other projectile travels through the air. Gun developers have typically tried to build weapons that fire with as "flat" or straight a trajectory as possible.

Trigger. The device on a firearm that ultimately causes it to discharge when pulled by a finger or otherwise engaged.

Trigger guard. A strip or band of metal that is fixed around the trigger in order to protect it from being damaged or accidentally pulled.

Wheel lock. An early modern ignition system that used a piece of flint set against a spinning metal wheel to set off a powder charge.

Windage. This refers to the difference between a projectile's diameter and the bore of a barrel. Expanding gases can slip through this gap past the bullet, similar to the co-viative principle. The term can also refer to the effect of wind on a projectile's trajectory.

Bibliography

Albaugh, William, III, and Edward Simmons. *Confederate Arms.* Harrisburg, PA: Stackpole, 1957.

Anderson, Jervis. *Guns in American Life.* New York: Random House, 1984.

Armstrong, David. *Bullets and Bureaucrats: The Machine Gun and the United States Army, 1861–1916.* Westport, CT: Greenwood, 1982.

"Arquebuse and Matchlock Musket Page." http://www.geocities.com/Yosemite/Campground/8551/ (accessed February 21, 2004).

Ayalon, David. *Gunpowder and Firearms in the Mameluk Kingdom: A Challenge to Mediaeval Society.* London: Valentine and Mitchell, 1956.

Barker, A. J., and John Walter. *Russian Infantry Weapons of World War II.* New York: Arco Publishing Company, 1971.

Batchelor, Peter, and Keith Krause. *2003 Small Arms Survey.* Oxford: Oxford University Press, 2003.

Black, Jeremy. *European Warfare, 1660–1815.* New Haven, CT: Yale University Press, 1994.

Blackmore, Howard. *Firearms.* London: Dutton Vista, 1964.

Boothroyd, Geoffrey. *The Handgun,* part 2. New York: Crown Publishers, 1970.

Bradley, Joseph. *Guns for the Tsar: American Technology and the Small Arms Industry in Nineteenth-Century Russia.* De Kalb: Northern Illinois University Press, 1990.

Brophy, Williams S. *The Krag Rifle.* Highland Park, NJ: Gun Room Press, 1985.

Brown, M. L. *Firearms in Colonial America*. Washington, DC: Smithsonian Institution Press, 1980.

Butler, David. *United States Firearms: The First Century*. New York: Winchester Press, 1971.

"C93 Borchardt Accessories." http://www.landofborchardt.com/C93_acc.html# MA (accessed February 21, 2004).

Chapel, Charles. *Guns of the Old West*. New York: Coward-McCann, 1961.

Childs, John. *Warfare in the Seventeenth Century*. London: Cassell, 2001.

Chinn, George. *The Machine Gun: Design and Analysis of Automatic Firing Weapons Systems,* vol. 4. Washington, DC: U.S. Government Printing Office, 1956.

Crosby, Alfred. *The Columbian Exchange*. Westport, CT: Greenwood Press, 1972.

———. *Throwing Fire: Projectile Technology through History*. New York: Cambridge University Press, 2002.

Croxall, Ian. "Snider Rifle 1867." http://www.britishempire.co.uk/forces/army armaments/rifles/sniderhistory.htm (accessed February 21, 2004).

Davis, William. *The Fighting Men of the American Civil War*. New York: Smithmark, 1991.

Diamond, Jared. *Guns, Germs, and Steel*. New York: Norton, 1997.

Diaz, Bernal. *The Conquest of New Spain*. Translated by J. M. Cohen. Harmondsworth, UK: Penguin Books, 1963.

Doyon, Keith. "Military Rifles in the Age of Transistion." http://militaryrifles.com/ (accessed February 21, 2004).

Dupuy, Trevor. *The Evolution of Weapons and Warfare*. Indianapolis, IN: Bobbs-Merrill, 1980.

Edwards, William. *Civil War Guns*. Harrisburg, PA: Stackpole, 1962.

———. *The Story of Colt's Revolver*. Harrisburg, PA: Stackpole, 1953.

Ellacott, S. E. *Guns*. New York: Roy Publishers, 1966.

Ellis, John. *Cavalry: The History of Mounted Warfare*. New York: G. P. Putnam's Sons, 1978.

———. *The Social History of the Machine Gun*. Baltimore: Johns Hopkins Press, 1986.

"Firearms: Muskets, Rifles and Carbines." http://www.researchpress.co.ukfirearms/ firearms.htm (accessed February 21, 2004).

Freemantle, T. F. *Evolution of Guns and Rifles*. Washington, DC: Arkansas, 1965.

Fuller, Claud. *The Breechloader in Service 1816–1917: A History of All Standard and Experimental U.S. Breechloading and Magazine Shoulder Arms*. New Milford, CT: Flayderman and Company, 1965.

Given, Brian. *A Most Pernicious Thing: Gun Trading and Native Warfare in the Early Contact Period*. Ottawa: Carleton University Press, 1994.

Gluckman, Arcadi. *United States Muskets, Rifles, and Carbines*. Buffalo, NY: Otto Ulbrich Company, 1948.

Gorshkov, Nikolai. "Russian Producer Wins Kalashnikov Rights." *BBC News* world edition, June 2, 2001. http://news.bbc.co.uk/2/hi/europe/2021173.stm (accessed February 22, 2004).

Greener, W. W. *The Gun and Its Development.* New York: Bonanza Books, 1967.

"Gunpowder and Weapons of the Late Fifteenth Century." http://xenophongroup
.com/montjoie/gp_wpns.htm (accessed February 21, 2004).

Guns and Gunfighters. New York: Bonanza Books, 1982.

"Guns of the Austrian Firm Steyr." http://trans.voila.fr/ano?anolg=65544&
anourl=http%3A//users.skynet.be/HL-Editions/roth/roth1.htm (accessed
February 21, 2004).

Hallahan, William. *Misfire: The History of How America's Small Arms Have Failed Our
Military.* New York: Scribner's, 1994.

Hamilton, Douglas. *Cartridge Manufacture.* New York: Industrial Press, 1916.

"Handgonnes and Matchlocks: A Preliminary Essay in the History of Firearms to
1500." http://homepages.ihug.com.au/~dispater/handgonnes.htm (accessed
February 21, 2004).

Hardy, Robert. *Longbow: A Social and Military History.* London: Bois d'Arc, 1992.

Hatch, Alden. *Remington Arms in American History.* New York: Rinehart and Company, 1956.

Haven, Charles, and Frank Belden. *A History of the Colt Revolver.* New York: Bonanza
Books, 1940.

Headrick, Daniel. *The Tools of Empire: Technology and European Imperialism in the
Nineteenth Century.* New York: Oxford University Press, 1981.

Held, Robert. *The Age of Firearms.* Northfield, IL: Gun Digest, 1970.

Helmer, William. *The Gun That Made the Twenties Roar.* London: MacMillan Company, 1969.

Hicks, James. *U.S. Firearms 1776–1956: Notes on Ordnance,* vol. 1. Beverly Hills,
CA: Fadco Publishing Company, 1957.

———. *What the Citizen Should Know about Our Arms and Weapons.* New York:
W. W. Norton, 1941.

Hobart, F.W.A., ed. *Jane's Infantry Weapons 1975.* London: Jane's Yearbooks,
1975.

Hogg, Ian. *The Story of the Gun.* New York: St. Martin's Press, 1996.

Hughes, B. P. *Firepower: Weapons Effectiveness on the Battlefield.* New York: Sarpedon,
1997.

Huntington, Roy. *Hall's Breechloaders.* York, PA: George Shumway, 1972.

Kaiser, Robert. "The Medieval English Longbow." *Journal of the Society of Archer-
Antiquaries* 23 (1980). http://www.student.utwente.nl/~sagi/artikel/long
bow/longbow.html (accessed February 22, 2004).

Keegan, John. *A History of Warfare.* New York: Vintage, 1994.

Keen, Maurice, ed. *Medieval Warfare: A History.* New York: Oxford University Press,
1999.

Kennedy, Paul. *The Rise and Fall of the Great Powers: Economic Change and Military
Conflict from 1500 to 2000.* New York: Random House, 1987.

Lazenby, David. "Cannons: That Diabolic Instrument of War." 1999. http://www.
middelaldercentret.dk/english/cannon2.htm#eksemplarer (accessed February 21, 2004).

Lenk, Torsten. *The Flintlock: Its Origin and Development*. Translated by G. A. Urquart. Edited by J. F. Hayward. New York: Bramhall House, 1965.

Leseman, Jeff. "History and Development of the M1911/M1911A1 Pistol." http://www.sightm1911.com/lib/history/hist_dev.htm (accessed February 21, 2004).

Lugs, Jaroslav. *A History of Shooting*. Feltham, UK: Spring Books, 1968.

"M16A2 5.56mm Semiautomatic Rifle." http://www.fas.org/man/dod-101/sys/land/m16.htm (accessed February 22, 2004).

Machiavelli, Niccòlo. *The Seven Books on the Art of War*. 1520. Transcribed into HTML by Steven Thomas for the University of Adelaide Library, 2003. http://etext.library.adelaide.edu.au/m/m149a/ (accessed February 23, 2004).

"Machinepistole 18, I." http://www.cruffler.com/historic-july00.html (accessed February 22, 2004).

Marcot, Roy. *Spencer Repeating Arms*. Irvine, CA: Northwood Heritage Press, 1983.

Martin, Paul. *Armour and Weapons*. London: Herbert Jenkins, 1967.

McNeil, William H. *The Pursuit of Power*. Chicago: University of Chicago Press, 1982.

Neal, W. Keith, and D.H.L. Back. *The Mantons: Gunmakers*. New York: Walker and Company, 1966.

Needham, Joseph. *Science and Civilization in China,* vol. 5. New York: Cambridge University Press, 1986.

Newman, James. *The Tools of War*. New York: Doubleday, Duran and Co., 1942.

North, Anthony, and Ian Hogg. *The Book of Guns and Gunsmiths*. London: Quarto, 1977.

O'Connor, Jack. *The Rifle Book*. New York: Alfred Knopf, 1964.

Owen, J.I.H., ed. *Brassey's Infantry Weapons of the World, 1950–1975*. New York: Bonanza, 1975.

Pacey, Arnold. *Technology in World Civilization*. Cambridge, MA: MIT Press, 1990.

Parker, Geoffrey. *The Cambridge Illustrated History of Warfare*. New York: Cambridge University Press, 1995.

———. *The Military Revolution*. New York: Cambridge University Press, 1988.

Partington, J. R. *A History of Greek Fire and Gunpowder*. Baltimore, MD: Johns Hopkins University Press, 1999.

Perrin, Noel. *Giving up the Gun: Japan's Reversion to the Sword*. Boulder, CO: Shambhala, 1980.

Peterson, Harold. *Pageant of the Gun*. Garden City, NY: Doubleday, 1967.

Peterson, Harold, and Robert Elman. *The Great Guns*. New York: Grosset and Dunlap, 1971.

"Photogallery of World War 2, Vapenmenu." http://ww2photo.mimerswell.com/ (accessed February 22, 2004).

Pope, Dudley. *The Great Gunsmiths*. New York: Spring Books, 1969.

———. *Guns*. New York: Spring Books, 1969.

Popenker, Max. "Modern Firearms and Ammunition." http://world.guns.ru/main-e.htm (accessed February 21, 2004).

"Pyrotechnics, Explosives and Fireworks." http://www.vectorsite.net/ttpyro.html (accessed February 21, 2004).

Reid, William. *Arms through the Ages.* New York: Harper and Row, 1976.

————. *The Lore of Arms: A Concise History of Weaponry.* New York: Facts on File, 1984.

"REME: The Corps of the Royal Electrical and Mechanical Engineers Museum of Technology Weapons Collection." http://www.rememuseum.org.uk/arms/armindex.htm (accessed February 21, 2004).

Ross, Steven. *From Flintlock to Rifle.* London: Associated University Press, 1979.

Ruffell, Wally. "The Gun: Rifled Ordnance." 1997. http://riv.co.nz/rnza/hist/gun/rifled1.htm (accessed January 28, 2004).

Russell, Carl. *Guns on the Early Frontiers.* Berkeley: University of California Press, 1957.

Shepard, G. A. *A History of War and Weapons, 1660 to 1918.* New York: Thomas Crowell, 1972.

"The Smith and Wesson Story." http://www.smith-wesson.com/custsupport/story.htm (accessed February 21, 2004).

Smith, Graham, ed. *Military Small Arms.* London: A Salamander Book, 1996.

Smith, Merritt Roe. *Harpers Ferry Armory and the New Technology: The Challenge of Change.* Ithaca, NY: Cornell University Press, 1977.

Sumrall, Al. "The Colt Model 1895 Automatic Machine Gun." http://www.spanamwar.com/Coltmachinegun.htm (accessed February 21, 2004).

Tallet, Frank. *War and Society in Early Modern Europe.* New York: Routledge, 1992.

Taylor, Chuck. "The M-1 Garand." *SWAT Magazine* (November 1982). http://www.pattonhq.com/garand.html (accessed February 22, 2004).

Thornton, John. *Warfare in Atlantic Africa, 1500–1800.* London: UCL Press, 1999.

Trenk, Richard. "The Plevna Delay: Winchesters and Peabody-Martinis in the Russo-Turkish War." *Man at Arms Magazine* (August 1997). http://www.militaryrifles.com/Turkey/Plevna/ThePlevnaDelay.html (accessed February 21, 2004).

Wahl, Paul, and Donald Toppel. *The Gatling Gun.* New York: Arco Publishing Company, 1965.

Warder, Bill, and Jill Loux. "History of Armor and Weapons Relevant to Jamestown." 1995. http://www.nps.gov/colo/Jthanout/HisArmur.html (accessed February 21, 2004).

Wilkinson, Frederick. *Firearms.* London: Camden House Books, n.d.

————. *Flintlock Pistols: An Illustrated Reference Guide to Flintlock Pistols from the 17th to the 19th Century.* London: Arms and Armour Press, 1976.

Williamson, Harold. *Winchester: The Gun That Won the West.* Washington, DC: Combat Forces Press, 1952.

Young, Peter. *The Fighting Man: From Alexander the Great's Army to the Present Day.* New York: Rutledge Press, 1981.

Index

About the Author

ROGER PAULY teaches history at the University of Central Arkansas. His interests include Victorian Britain, military history, and the history of science and technology. He earned his Ph.D. from the University of Delaware.